Watts Pocket Handbook

Twenty-ninth edition

Edited by Trevor Rushton

Routledge
Taylor & Francis Group

LONDON AND NEW YORK

Twenty-ninth edition published 2016
by Routledge
2 Park Square, Milton Park, Abingdon, Oxon OX14 4RN

and by Routledge
711 Third Avenue, New York, NY 10017

Routledge is an imprint of the Taylor & Francis Group, an informa business

First edition published 1983
Twenty-eighth edition published 2012

British Library Cataloguing-in-Publication Data
A catalogue record for this book is available from the British Library

Library of Congress Cataloging in Publication Data
A catalogue record has been requested for this title

ISBN: 978-1-138-66595-8 (pbk)
ISBN: 978-1-315-61958-3 (ebk)

Typeset in Frutiger 10/12 pt
by Fakenham Prepress Solutions, Fakenham, Norfolk NR21 8NN

Printed and bound in Great Britain by
T International Ltd, Padstow, Cornwall

Contents

Watts Pocket Handbook

Back in print for the first time in years, the *Watts Pocket Handbook* renews its commitment to share industry knowledge by providing technical and legal information across a comprehensive spread of property and construction topics.

Compiled by the Watts Technical Director, the *Handbook* provides specialist information and guidance on a vast selection of construction-related subjects including:

- contracts and procurement
- insurance
- materials and defects
- environmental and sustainability issues.

Watts Pocket Handbook remains the must-have reference book for professionals and students engaged in construction, building surveying, service engineering, property development, and much more.

Trevor Rushton is Technical Director at Watts Group Limited. He has been the editor of the *Watts Pocket Handbook* for over 15 years and involved since the first edition in 1983. He is the author of *Deleterious and Hazardous Building Materials* and *Defects in Industrial and Commercial Buildings* both published by RICS Books together with numerous articles for the *Building Surveying Journal*. He is a regular speaker at many RICS conferences and seminars.

Introduction

The first edition of the *Watts Pocket Handbook* was published in 1983. Inspired by *Hurst's Architectural Surveyors' Hand Book* of 1864, we set out to provide a collection of checklists and information in précis form to aid the practitioner, and the volume ran to a mere 61 pages.

In 2013 – in line with the revolution in digital publishing – we took the handbook online with the expectation that this bold experiment would enable us to expand upon the content and enable us to provide more images, tables and other useful data.

However, it soon became apparent that readers preferred the look and feel of a hard copy. We know that the *Watts Pocket Handbook* has carved a reputation as an indispensable point of reference: countless surveyors have referred to the handbook as being an essential revision aid for the APC, and many old friends proudly display their dog-eared copies.

We always listen to our clients, and I am therefore delighted to introduce the 29th edition of the *Watts Pocket Handbook* in hard copy. We've updated all sections and slimmed the book down a little to restore its pocket status. We hope that you will find it as informative as ever, and do please send your comments and suggestions for future editions.

1
Property investment and ownership

1.1 Technical due diligence

Best practice in building surveys
Trevor Rushton

In principle, there is little difference between the inspection and method of reporting upon commercial and industrial property compared with surveys of residential or other property. Many of the principles of setting up, confirming the instruction, and report writing are identical, as is of course the process of reflective thought that is a key ingredient in any survey for investment or occupation.

The form of construction and style of building is often different (although not exclusively so), and surveyors must be confident that they are suitably experienced to deal with the property and nature of the instruction; it is far better to decline an instruction if the scope of the survey would be beyond your expertise than to muddle through and potentially miss an important issue.

Commercial and industrial property

Commercial and industrial buildings take many forms, but just as importantly, purchasers and occupiers have many different requirements. The form of reporting on an identical building for an investor could require a different emphasis from that needed for an occupier, or, indeed, that needed to be included in a vendor's pack.

For Chartered Surveyors, appropriate guidance may be found in *Building surveys and technical due diligence of commercial property* (hereafter referred to as the 'Commercial/Industrial GN'). At the time of going to press a new Practice Statement is in preparation. The 2010 edition provides comprehensive guidance and advice for surveyors and engineers involved in building surveys of commercial property. It is written to apply in England and Wales, although its content is equally applicable elsewhere. RICS members are not obliged to follow the advice and recommendations given and are free to make their own mind up when it comes to what to include or how to present the information, and, as long as the advice given is properly thought through with the correct level of skill and care, there should be no need for concern. However, in cases where professional negligence is being considered, adherence to the guidance note would at least give partial defence by virtue of having followed recognised practice.

Taking instructions

Managing client expectation is the key to ensuring that the client is satisfied with the service and the quality of the report. The instruction stage is often botched or not recorded properly – in many cases this is where a job will start to go wrong. It is very important to take the time to understand the client's exact requirements, the purpose of the survey, the scope and limitations, specialist inspections, etc.

The RICS Guidance Note *Taking instructions* sets out in some detail the issues to be resolved at this stage, including:

- the identity of the client (this is sometimes unknown until completion);
- the scope of service;
- timescales;
- the fee;
- terms and conditions of business; and
- limitations.

At this stage it will be apparent whether or not specialist subconsultants will be required. Commonly this will include mechanical, electrical, public health, and lift engineers, as well as cladding consultants and environmental or structural engineers. Are these consultants to be paid direct by the client (preferable) or by the surveyor? In the latter case, make sure that these appointments are tied down – preferably with back-to-back agreements – and that Professional Indemnity Insurance (PII) levels are matched.

Time pressures are usually acute, and purchasers often make optimistic promises in order to secure preferred-bidder status. This often places unreasonable pressure on the surveyor. Whereas at one time quick 'walk-around' surveys were frowned upon, the Commercial/Industrial GN acknowledges that it may be necessary to accept this type of instruction, but points out that surveyors must ensure that the client is under no illusion that compromising on time will reduce the quality of the report. Under these circumstances you should take great care to explain the limitations of such an approach. Preferably, demands to compromise should be resisted, but, as an alternative, spend the same time on site and reduce the time reporting by focusing on key issues only. Composing a short report is as much of an art as writing a full one; in many respects the task is more difficult.

There is a risk of allowing quality standards to drop in favour of fiercely competitive fee quotations and timescales that are unrealistic. However, a brief report need not be a 'light-weight' report if it is prepared properly and based upon proper consideration of the evidence.

Health and safety

'Time spent in reconnaissance is seldom wasted' is a maxim that surveyors should remember. Aside from giving the opportunity to plan resources, assess the fee, and determine the need for specialist investigations, it also enables time for a proper risk assessment to be prepared.

Regrettably, too few surveyors treat health and safety as seriously as they should; these issues fall under criminal law, and penalties for non-compliance are severe. Employers have a duty to protect the safety of their staff – one of the key issues being a proper risk assessment. The risk assessment should be appropriate for the particular circumstances, but it should be recorded in writing.

Key points on health and safety issues are:

- risk assessment – can be in any appropriate form, but it should be recorded and filed;
- the employer has a duty not to put the surveyor at risk;
- the surveyor must bring limitations to the employer's notice;
- wear appropriate dress; and
- use appropriate equipment and maintain it in good order, ensuring that:
  - the places in which you work are safe;
  - safe working practices are clearly defined;
  - first aid equipment is available;
  - you do not endanger yourself or others;
  - you protect anyone who uses your services;
  - you protect the people that work for you; and
  - you protect anyone affected by your work.

Perfect planning prevents poor performance – make sure that your team is properly briefed as to their roles and responsibilities, the purpose of the survey, and the extent of the instruction. Furthermore, it is vital to arrange proper access with the tenant/occupier.

Preparation is essential; it gives the opportunity to seek out relevant documents, for example, as-built drawings, certificates, previous reports, and the like. It also establishes any particular working practices on site that could impinge upon the survey and highlights any particular health and safety issues that may be relevant. Arranging for a hydraulic platform to attend site to facilitate a roof inspection is fine, but not if the tenant is not expecting it and has, say, a 24-hour operation with lorries coming and going all the time. Frequently, working arrangements on large industrial or distribution sites are such that the operators demand full risk assessments before using access equipment. Trying to sort these out at 10:00 on the morning of the survey with the team on site is not the most productive use of time. (See also *Surveying Safely* published by RICS.)

Equipment and note taking

Unlike the equivalent residential survey guidance, the Commercial/Industrial GN is not specific as to the equipment that you must use when conducting a survey – it merely lists a range of basics that a survey kit 'may' include. Clearly, one should expect to take the equipment that would be necessary to undertake the task properly. It would not be very sensible, for example, to turn up without a measuring tape if you had specifically been asked to prepare a reinstatement valuation.

Many surveyors prefer to use digital recording equipment to take comprehensive notes, which there is no problem with, but:

• beware of things like wind and background noise;
• keep checking to make sure that you are actually recording;
• beware that sometimes you can record things you don't intend; and
• consider whether it is productive to have secretarial staff typing notes all day before you can write the report.

The key issues are to take good, legible notes and to retain these on file. Do not expect to dictate the final version of the report on site – this does not allow time for reflective thought.

The inspection

Different survey types demand different levels of inspection, and this should be negotiated with the client at the outset. In the event that physical restrictions are more onerous than expected, one must communicate this back to the client as soon as possible, and certainly not be left to a short line in the completed report.

Making friends with the occupier is a good move; often there will be issues with site security and the need to be accompanied (although this can often be worked around by diplomacy – and weight of numbers). A team of five or six cannot do their job properly if one security guard has been allocated, but patience and compromise can often work well in achieving proper access.

Surveyors will each have their own methodology for dealing with surveys, and the guidance notes do not dictate how inspections should be carried out. However, a methodical approach is the key, and this typically means that an elemental approach is preferred.

Conducting a good building inspection and preparing a good report should not be treated lightly. Taking a graduate with you and then asking him or her to 'do the insides' while you examine the exterior – or the interesting parts – is not very good training, and certainly runs the risk that small pieces of evidence that relate to a bigger picture may not be assimilated. Technical knowledge is all too often taken for granted – but, if you don't know what can go wrong, how can you possibly be expected to look for it? It is very easy to look at something and to describe it, but does this necessarily address the real issues?

Surveyors are not infallible, but it is imperative that you do not fall into the trap of simply describing what you see without thinking about the job that the element is doing and how it interrelates with other components or elements. One must think behind the surface, consider the building in its underpants.

The guidance note sets out many of the types of issues that ought to be identified during an inspection. Some of the more important issues are summarised below:

- In a large building, it is not anticipated that the surveyor will check each and every unit/window or feature where there is a lot of repetition.
- Comment on fire resistance, compartmentation, thermal, and insulation standards.
- Assess the effectiveness and condition of structural frames.
- Check the degree of flatness along with surface defects (particularly important for industrial and storage buildings).
- Inspect ceiling and floor voids. Check space for services/ crossovers as well as fixity of pedestals/suspension systems.
- Inspect roof voids where they are accessible.
- Efforts should be made to inspect ducts and similar enclosed areas where this is possible.
- Check staircase width, handrail height, etc. and general compliance with Building Regulations.
- Check provision of sanitary facilities for DDA and Workplace Regulations.
- Disability considerations should include different textures to wall and floor finishes.

Services installations

In complex buildings, mechanical, electrical, and public health (MEP) installations are usually inspected by specialists, often under the direction of the surveyor. However, in certain circumstances, this will not occur, and a surveyor may be commissioned to provide general advice. The object here is to carry out a general visual inspection and to form a view as to the type of installation, the materials used, and the need for more detailed investigations. RICS cautions against direct appointments and payment of subconsultant's fees for fear of assuming a liability for parts of the report.

One point to watch is the services engineers' tendency to rely upon Chartered Institution of Building Services Engineers (CIBSE) tables of life expectancy. Too frequently we find observations like 'the installation is 25 years old and therefore at the end of its life' without comment upon the standard of maintenance and use of the particular component, when in reality a longer life can be anticipated. It is the surveyor's role to challenge such observations and make sure that the advice given is sensible and practicable.

Surveyors will be expected to provide general advice on issues such as compliance with relevant lighting standards under the Workplace Regulations. The position of gas service intakes as well as the condition of appliances and pipe runs should also be identified. Any smell of gas should be reported immediately.

Surveyors are advised to note and report on the nature of the installation including pipework, point of entry, stopcocks, and the like, as well as compliance with water by-laws, visual corrosion, and leaks. The guidance note suggests a represent-ative number of taps be turned on.

Hot water and heating installations should be checked for operation and general condition.

Underground drainage is often an area that fails to be mentioned properly – either through lack of access for visual inspection (reasonable) or because the surveyor forgot to bring drain cover keys or could not be bothered to lift light-weight covers (unreasonable). Be clear about the extent of the inspection and whether or not covers are to be lifted. The guidance note advises surveyors to:

- open all reasonably accessible lightweight inspection chamber covers within the curtilage of the property;
- record assumed drainage runs; and
- report their general condition.

CCTV inspections might be commissioned, but beware of the limitations of these – it is not always possible, within the time available, to inspect and plot all runs. Furthermore, silting and encrustation may mean that an element of jetting and cleaning is needed before a survey can be undertaken. It is worth being very clear about the scope of work in these circumstances and agreeing this with the client at the outset. For example, if a fixed sum is required, agree that this be limited to what can be achieved with this equipment for a fixed time.

Another point worth checking is the direction that drainage runs take when they leave site; for example, do they cross adjoining land?

External areas

The scope of the external inspection depends upon instructions, but remember to be alert for boundary issues, tree growth, Japanese Knotweed, Giant Hogweed, power lines, and the like, as well as any particular topographical points of concern.

Health and safety legislation

A detailed analysis of the property for compliance is not normally undertaken unless specifically agreed with the client at the outset. Surveyors should be aware of relevant legislation that might affect health and safety in the building and advise accordingly. The following issues are particularly important:

- protection against falling;
- sanitary provision;
- lighting;
- floors and traffic routes;
- glazing;

- working at height;
- fire precautions – resistance, means of escape and protection; and
- ventilation and cleanliness.

Security

With increased terrorist threats, security issues have become more important and should command some attention to detail in an inspection and report.

One of the points recommended for attention is the existence and condition of window film. A word of caution though: to be effective, the film must transfer explosion loads back to the structure of a building where it is more likely that the loads can be dissipated safely by the inertia of the building. This can only be achieved with physical connection to glazing beads with silicone at the perimeter. Furthermore, providing film on laminated glass alters the properties of the glass and can give rise to a greater propensity for the glass to rupture around the edge – and thus form a large and lethal projectile.

The guidance note recommends reference to guides like *Secured by design*, which may help a surveyor comment on security aspects.

Within the UK, crime mapping has now become established and via local police websites the surveyor can interrogate data down to sub-ward level. Such material can provide useful background information when assessing risks.

Social inclusion

Unless otherwise instructed, a surveyor would not be expected to carry out a detailed audit of the premises. However, the identification of key barriers to access and any other significant points of concern ought to be mentioned, not forgetting that there is much more to social inclusion than simply the provision of ramps and disabled toilets.

Environmental issues and sustainability

Investors and funders are sensitive to environmental issues. More often than not an investment purchase will involve the preparation of a suitable Phase I (and sometimes a Phase II) environmental audit by appropriately qualified specialists. In these circumstances (and assuming that the process includes a site visit), it is important to establish exactly where responsibility stops to avoid duplication and, more importantly, conflicting information.

If an audit is not being undertaken, the surveyor should be alert to the usual environmental risks and comment accordingly. In the UK, several valuable sources of information (such as the Environment Agency and the Scottish Environment Protection Agency) have accessible postcode-search facilities to allow detailed advice on matters such as flooding risk. However, take care to understand the limitations of the material that is provided and in particular what is meant by the risk assessments themselves.

Knowledge of gas protection measures – membranes and passive venting – is also useful. Sometimes solicitors turn up a requirement for gas protection in planning documents or past environmental reports, and it will be useful to be able to reply that indeed measures were in place.

Sustainability should now be a key component of a building survey. RICS publishes guidance on sustainability with reference to the property lifecycle. Examination of that guidance will reveal that many of the topics listed above fall neatly into the sustainability heading – it is not simply a matter of carbon reduction and 'green' materials, but more an all-encompassing assessment of the building and its impact on the people who use it, as well as its impact on the environment. Matters such as biodiversity, water management, waste management, health, and welfare are all relevant and should be considered.

The *impact* of orientation (not just the orientation) of the building should be clearly stated, as this affects the performance of materials used in the construction of the building as much as thermal performance.

With the impact of climate change becoming ever more topical, it is to be expected that surveyors will need to take a more active role in the analysis of the performance of buildings. At present, surveyors are advised to:

- describe the thermal shell, taking into account orientation;
- consider the nature of the heating and cooling systems;
- consider the nature of artificial lighting; and
- provide advice on practical methods of upgrading insulation and measures to reduce associated condensation risk.

In many circumstances, an Energy Performance Certificate (EPC) is required and should be presented by the vendor. It is worth checking to see if the certificate is in place before the inspection to find out whether additional work is likely to be required. Data collection for an EPC needs to be methodical and careful. Knowledge of the modelling systems is useful, as most systems rely upon simplified models as opposed to detailed CAD drawings of the building. There can be a tendency to collect too much data if you are unfamiliar with the procedure. Refer to www.ndepcregister.com for details of non-domestic EPCs (this is a searchable database).

Noise and disturbance issues to note would include:

- the effects of noise from external sources;
- sound insulation of party structures; and
- other possible nuisances.

While testing for deleterious materials may be outside the scope of the survey (at least to begin with), surveyors should be aware of the key issues and materials involved, and provide reasoned advice accordingly. Typically, the list of materials normally considered would include:

- high alumina cement (HAC);
- calcium chloride additive;
- asbestos;
- lead in plumbing and paint;
- woodwool as permanent shuttering;
- toughened glass, particularly in overhead situations; and
- mundic (mainly in the southwest of England).

Some clients may have particular concerns with, for example, glass reinforced concrete (GRC), reinforced autoclaved aerated concrete (RAAC) planks, or machine made mineral fibre (MMMF). In Scandinavian countries, the use of mastic containing polychlorinated biphenyls (PCBs) has been viewed with concern, and owners can go to great lengths to decontaminate their buildings of this material.

For buildings constructed up until 1974, it is common to consider tests for high alumina cement (HAC). Calcium chloride was effectively banned from 1978, so buildings constructed later than this should be free of those materials in construction (although, of course, chloride contamination can still occur as a result of atmospheric conditions or exposure to de-icing salts).

Asbestos merits a specific section in the report owing to the special considerations that this material warrants. Ask to see a copy of the asbestos register and consider its adequacy.

For industrial and retail buildings, particularly those constructed in the last 15 years or so, the likelihood is that composite cladding materials will have been used. Since 2000, insurers have been wary of panels that contain combustible cores – expanded polystyrene (EPS), polyurethane (PUR), and some types of polyisocyanurate (PIR [or polyiso, or ISO]). Of these, EPS is of most concern, as this will be found in cold stores and self-supporting panels. Some insurers may significantly increase premiums or deductibles where there is uncertainty as to the nature of the core and whether or not it is Loss Prevention Standard (LPS) approved.

While the industry has taken a slightly more robust view of late, it is important to take steps to try and identify the nature of the core and if it is of an approved type. If it is not, provide advice on the thickness and location and particular fire risks – arson, cardboard storage, battery charging, and the like-to facilitate a proper risk assessment.

Warning of the existence of deleterious and hazardous materials is one thing, but their existence needs to be placed in perspective and straightforward advice needs to be given. For more information, see *Investigating hazardous and deleterious building materials*, published by RICS Books.

Further enquiries

During the course of the inspection, various technical and legal issues may arise. It is appropriate to draw attention to these points with the client and the client's legal adviser. However, questions and referrals should be relevant and sensible – not issues that are readily within the ambit of a surveyor to establish (for example, contacting the manufacturer of a composite panel to find out if it complies with Loss Prevention Standards).

Recommendations to do something inappropriate – like check the warranty package on a 12-year-old building – may cast the surveyor in poor light and should be considered carefully rather than added as a 'throw-away' line.

Always make it clear whether or not the enquiries and investigations need to be concluded before proceeding to completion.

Reporting

The reporting stage is as important as the data-collection stage. Your overriding objective will be to affect your reader precisely as you would wish, which means that you must take care to compile the report carefully; satisfy the brief and make sure that you have given sufficient thought to the matter. It is fairly easy to describe buildings; the real key is to understand the various pieces of evidence that have been gathered, weigh them up, and advise accordingly.

Use plain English in reports: avoid jargon and the excessive use of acronyms. Keep the text active and try and avoid long, passive sentences. Ask the following questions:

- What is it?
- What is wrong with it?
- What needs to happen to put it right? Are there options?
- What happens if you don't put it right?
- Who pays for it?
- Above all – what is your recommendation?

Further information
Building surveys and technical due diligence of commercial property (4th edition), RICS guidance note, 2010. Retrieved from www.rics.org/uk/knowledge/professional-guidance/guiance-notes/building-surveys-and-technical-due-diligence-of-commercial-property/.

Surveys of residential property (3rd edition), RICS guidance note, 2013. Retrieved from www.rics.org/uk/knowledge/professional-guidance/guidance-notes/surveys-of-residential-property-3rd-edition/.

Identifying the age of buildings
Michael Wright

This section explains where to look when seeking to assess the age of a building and fully and properly understand its forms and materials of construction. Almost without exception, any building, unless of very recent construction, will, at some time in its life, have undergone some form of change, modernisation, or conversion that may well hide the age and materials of the original construction.

Archival research

It is a well-established principle of good conservation practice, as advocated by English Heritage, that "understanding the significance of places is vital ... in order to identify the significance of a place, it is necessary first to understand its fabric and how and why it has changed over time".

As a result, archival research is generally considered to be a key process when seeking to establish both a date for an existing building and dates for any past alterations made to the building in question. Indeed, it is important to impress upon the client the importance and potential value of such research. Even if the building is not a historic building, some records of the original date of construction will exist somewhere in the files of the local administration.

If the building, or 'heritage asset', is listed as being of Special Architectural or Historic Interest, or a Scheduled Ancient Monument for that matter, the listing description or entry in the Scheduled Monuments Register will provide some indication of the believed age of the building or monument. A word of caution here, however, as these estimated ages are based, in most cases, on external inspections only, and even the most experienced inspectors of historic buildings and ancient monuments have been known to have been deceived.

In the past, a more recent age has been mistakenly attributed to a structure based on external elevations that are the result of, say, a late Victorian refronting of a Georgian building, or an early 18th-century brick refronting of a Tudor or earlier timber-framed building. The latter is often referred to as a 'Queen Anne front on a Mary Anne back', and so the surveyor is always well advised to inspect the rear of the property as well as the front.

The heritage authorities are very gradually introducing 'Statements of Significance', either as stand-alone documents or integrated into more comprehensive listing descriptions, and the National Planning Policy Framework (March 2012) now states that "local planning authorities should require an applicant to describe the significance of any heritage assets affected", thereby making archival research all the more important, and potentially required when undertaking certain works to an existing building.

Where to go to find records

Every county in Britain has a County Archivist who can be contacted through the local authority of the area in which the building is situated. Alternatively, contact can be made via the relevant heritage authority – Historic Scotland, Historic England, Welsh Heritage (CADW), or the Northern Ireland Environment Agency. Alternatively, the local authority Conservation Officer may be able to assist.

Archives, be they county, district, or more local, even down to parish level, may include one or more of the following:

- maps, especially tithe maps, showing (down to considerable detail) ownerships and the building in outline on each site or plot;
- sale documents, especially auction details, that may give indications of believed age(s) of the building;
- newspaper and other articles indicating the believed age(s) of the building;
- deeds registers, which can be particularly useful in establishing exact dates for the original building lease from the lord of the manor or landed estate owner, granting the right to the construction of the building being considered;
- local authority building by-laws and drainage permissions or approvals for the construction of the building and subsequent alterations. This last one usually only reaches back to the mid-19th century, but the equivalent landed

estate or manorial records can reach back much further. If the building is still held in freehold or the equivalent by the landed estate or manorial estate, then such records may still be held by them. However, some such estate records have been transferred to the county or district archives and/or to the National Monuments Records Office in Swindon. The English Heritage website (www.english-heritage.org.uk) will link into 'Research & Conservation' and 'Learning & Resources' with various sites via 'Online Resources', including Images of England and Access to Archives. The United Kingdom strand of the A2A Archives database (at www.nationalarchives.gov.uk/a2a) links to over 400 record offices. English Heritage (EH) is working very hard to bring both their images of England and Heritage Gateway up to the level of data that are needed for at least basic 'Statements of Heritage Significance', and searching the EH website for both the Images of England Advanced Search (which you have to register to use,, but it is at present free to use and search on) and Heritage Gateway will give you the best available data online.

- Welsh Heritage (CADW) has 'Coflein' and other free to use, no need to register, databases on www.rcahmw.gov.uk/.
- For Scotland, see www.historic-scotland.gov.uk under the category 'Looking after our heritage' and subcategory 'Searching for a listed building'. This is a comprehensive integrated search system across all categories – from listed buildings and Scheduled Monuments through wreck sites, RCAHMW (Royal Commission on the Ancient and Historical Monuments of Wales) records, gardens, and landscapes to local authority historic environment records.
- For Northern Ireland see www.ni-environment.gov.uk/.

Published and unpublished archival research

The *Victoria county histories*, which should be available in the county or district central reference library, may contain a reference to your building. These are increasingly available online at www.british-history.ac.uk. Search by either 'local history' or by 'region'.

London has the *Survey of London* volumes now produced by the Survey of London branch of English Heritage. Those started in the late nineteenth century only cover a part of the historic areas of London and are available in principal reference libraries online at www.british-history.ac.uk (link to 'Survey of London'). The British History site has numerous additional online sources covering London and the UK, including records and documents back to the thirteenth century.

Local history librarians are a mine of information in such a search. They and the county or district archivists may also be able to assist in pointing you towards unpublished works held by them or produced by local historians.

Research in specialist public archives can also be extremely positive in producing plans and documentation. For any building that is or has been in government or Crown ownership, including the Crown Estates, the Public Record Office at Kew can be invaluable, with large parts of their catalogues available online. See www.nationalarchives.gov.uk and enquiry@nationalarchives.gov.uk.

For a building by an important architect, the V&A/RIBA Drawings Collection at the Victoria and Albert Museum can produce drawings back to the seventeenth century. See www.vam.ac.uk/collections/architecture/va_riba/index.html.

For London, the Metropolitan Archives (T: 020 7332 3820) hold papers dating back to the sixteenth century. See www.cityoflondon.gov.uk/things-to-do/london-metropolitan-archives/about/Pages/archive-service-accreditation.aspx.

For any building that has at any point in its life been in the direct ownership of (or occupation by) the Monarchy, the Royal Archives at Windsor Castle can produce vital details unobtainable elsewhere.

The building

Once the types, forms, and materials of construction used in particular periods have been determined, the less altered areas of the building can be very revealing. This is particularly true of roof spaces, basements, rear elevations, back or rear additions, or anywhere else that has escaped the previous owners' 'improvements'. 'Above-ground archaeology' is a common term for such on-the-property investigations.

What to look for – the ages of architecture and their clues

Remember that every building will have been altered and changed over the decades, so each is an amalgam of various periods. It is crucial to seek to ascertain in the first instance the original building date of the first part of the building in question.

For example, a major country mansion attributed in the listing description to Edmund Blore as of 1834–38 in Tudor style turns out to be a major Elizabethan house of the 1500s, badly damaged by fire in 1836, and rebuilt in Mock Gothic style incorporating the Elizabethan chimneys and cellars.

So what do we look for in the periods of architecture in Britain and where might we refer to for assistance? Knowledge gained over decades helps, but if you are starting out you might carry the following with you as an *aide-mémoire*.

The table below gives a brief description of the key ages of architecture in Britain. Even if the original windows and doors have been replaced, the form, shape, and materials and forms of construction of the openings (especially the head arch, lintel, bressumer, or beam over) is least likely to have changed.

Table 1.1 A brief list of the ages of architecture in Britain

Form of architecture	Approximate period	Characteristics
Romanesque	500 to 1200 AD	The spread of the classical architecture of Rome adapted to incorporate the heavy, stocky, squat Romanesque architecture characterised by rounded arches.
Gothic	1100 to 1450 AD	Pinnacles reaching for the sky, characterised by the creation of the great cathedrals.
Renaissance	1400 to 1600 AD	The return from Gothic to the classical precise rules of the 'Golden Section' of the Graeco-Roman architecture of 850 BC to 476 AD. Influenced by the 'Age of the Awakening and the Enlightenment' from the Grand Tours of Europe and the rediscovery of Roman and Greek architecture. Remember Andreas Palladio and Palladian architecture.
Baroque	1600 to 1830 AD	While in such places as Italy Baroque is characterised by opulent and dramatic churches with very extravagant decorations, Britain followed more of a restrained French form of highly ornamented Baroque style combined with classical features.
Rococo	1650 to 1790 AD	In essence the latter phase of Baroque, with graceful white buildings incorporating sweeping curves.
Georgian	1720 to 1800 AD	This stately symmetrical style predominates in British and Irish towns and cities, from London to Bath to Edinburgh to Dublin – the age of Jane Austen so frequently portrayed on screen.

Form of architecture	Approximate period	Characteristics
Neoclassical	1730 to 1925 AD	Renewed interest in Palladio and Palladianism brought a return to classical shapes and forms.
Greek Revival	1790 to 1850 AD	Major classical mansions and buildings featuring columns, pediments and details inspired by Greek architecture of 850 BC to 50 AD.
Victorian	1840 to 1900 AD	The Industrial Age with its greater use of iron brought many innovations in the use of new materials and a flurry of architectural styles – Gothic Revival, Italianate, Queen Anne and Romanesque, all using previous styles and sometimes intermixing them.
Arts and Crafts Movement	1860 to 1900 AD	The backlash against Victorian industrialisation, with renewed interest in simplistic handicraft forms being applied to architecture.
Art Nouveau	1890 to 1914 AD	Originally patterns in fabrics, this spread rapidly to architecture and is characterized by asymmetrical shapes, arches and decorations incorporating curved, often plant-inspired, designs.
Beaux Arts	1895 to 1925 AD	Also known as Classical Revival, this incorporates formal orders, symmetry, and elaborate ornamentation.
Neo-Gothic	1905 to 1930 AD	Here Medieval Gothic forms were applied to the new breed of tall office blocks and skyscrapers.
Art Deco	1925 to 1939 AD	The Jazz Age, when zigzag patterns and vertical lines created dramatic stage-like effects on facades.

Form of architecture	Approximate period	Characteristics
20th century extending into 21st century	1900 to the present day	Confusion of forms from Walter Gropius and Bauhaus through to Modernism, Brutalism (functionalist blocks of concrete and steel), and Post-Modernism, the last of which represents some return to classical and true forms of architecture. Styles collide and seek exuberance in building techniques and stylistic references, e.g. a skyscraper with a crowning oversized Chippendale pediment.

Further information

The best way to get used to identifying styles and ages and matters on site is to carry an easy-to-use, pocket-sized illustrated reference book such as:

The observer's book of architecture, Penoyre, J. and Ryan, M., Godfrey Cave Associates, Ltd, 1992; Number 13 in a series of most useful reference books (the best by far – out of print but may be available through second-hand booksellers/websites – ISBN-10: 1854710397, ISBN-13: 978-1854710390).

British architectural styles: An easy reference guide, Yorke, T., Countryside Books, May 2008 (ISBN 978-1846740824).

Listed buildings and conservation areas

Michael Wright

The United Kingdom enjoys a proud history of being at the forefront of developing the practice of conservation and restoration. Since William Morris issued his manifesto in 1877 for The Society for the Protection of Ancient Buildings, the UK has enacted legislation and governance, at both local and national levels, to promote the retention and protection of our heritage assets. The importance that our built heritage holds to our quality of life, our sense of place, and our understanding of our shared history is now widely recognised and advocated across both public and private bodies.

However, efforts are still having to be made to protect heritage assets from perceived notions of redundancy, whether that be in respect of economic factors or sustainability factors, and considerable research has been undertaken by public heritage bodies in recent years to defend and highlight the importance of the UK's rich heritage stock.

As an example, in 2015 Historic England issued their 'Heritage and the Economy' which sought to record the value to the UK economy in heritage properties. Some key headlines included that 'heritage tourism represents 2% of the UK's GDP in 2011', and that 'repair and maintenance on historic buildings directly generated £4.1 billion GDP in England in 2010'. In addition, it was recorded that the built heritage is a key part of the UK brand, with the UK being ranked fifth out of 50 nations in terms of being rich in historic buildings and monuments and seventh for cultural heritage in the Nation Brand Index.

In a response to the UK government's increasing focus on sustainability and the importance of addressing energy use in existing buildings (considered to account for nearly half of the UK's total carbon dioxide emissions[1]), Historic England have produced guidance stating that "there is no inherent conflict between the retention of older buildings and the principles of sustainability"[2] and have gone on to highlight that all pre-industrial buildings, when first built and inhabited, were, by definition sustainable, and made zero use of fossil carbon in either their construction or use, and also that thermal upgrading of existing buildings accords well with both conservation and sustainability principles.

In any event, the task set before both public and private bodies, to advise and inform the public and clients as to the importance and value of the heritage stock, continues. Some of the legislative tools available to local authority bodies to control proposed works to heritage assets are set out below.

Listed buildings

The effect of a building being listed is that it allows the local planning authority to review proposals for alterations to a listed building before works are carried out, and consent is granted if the proposal is deemed to be acceptable. Any proposals to demolish, alter, or extend internally or externally a listed building will require listed building consent. Listing does not necessarily mean that proposed alterations will be refused, however; it does generally mean that demolition will not be permissible.

The number of listing building entries (at the time of writing) is approximately:

- England: around 376,000 listed building entries
- Scotland: around 47,000 listed building entries
- Wales: around 30,000 listed building entries
- Northern Ireland: around 8,500 listed building entries

Governing legislation in respect of the listing process in the UK is as follows:

- England and Wales: Planning (Listed Buildings and Conservation Areas) Act 1990 and The Enterprise and Regulatory Reform Act (ERRA)
- Scotland: Planning (Listed Buildings and Conservation Areas) (Scotland) Act 1997
- Northern Ireland: The Planning (Listed Buildings) Regulations (Northern Ireland) 2015

Listing categories

Buildings throughout the UK are divided into three categories according to their relative importance. The categories of listing are not uniform across the UK, and are defined as below. It is important for the surveyor to be aware that all categories are advisory and have no statutory status, and that all categories are treated equally under the relevant legislation.

England and Wales

- Grade I: buildings of exceptional interest
- Grade II*: buildings of particular importance
- Grade II: buildings of special interest

Scotland

- Grade A: buildings of national or international importance, either architectural or historic; or fine, little-altered examples of some particular period, style, or building type
- Grade B: buildings of regional or more than local importance; or major examples of some particular period, style, or building type, which may have been altered
- Grade C: buildings of local importance; lesser examples of any period, style, or building type, as originally constructed or moderately altered; and simple, traditional buildings that group well with other listed buildings

Northern Ireland

- Grade A: buildings of special or national importance including both outstanding grand buildings and the fine, little-altered examples of some important style or date
- Grade B+: buildings of special importance that might have merited 'A' status but for relatively minor detracting features, such as impurities of design or lower quality

additions or alterations; also buildings that stand out above the general mass of grade-B1 buildings because of exceptional interiors or some other features
• Grade B1 and B2: buildings of more local importance or good examples of some period of style; some degree of alteration or imperfection may be acceptable

Conservation areas

'Conservation areas' have previously been defined as "areas of special architectural or historic interest, the character or appearance of which it is desirable to preserve or enhance"[3].

Conservation areas are designated by the local planning authority, which is required to determine which parts of its area are of special architectural or historic interest. The public is usually consulted on any proposals to designate conservation areas or to change boundaries.

When coordinating works to a property that is located within a conservation area, one should be aware of the following controls which are in place for these designated areas:

• Property Alterations: When alterations are proposed to a building within a conservation area (such as cladding, inserting windows, installing satellite dishes and solar panels, adding conservatories or other extensions, laying paving or building walls) you may need permission from the local authority for these works.
• Trees: If you are considering cutting down a tree or doing any pruning work, you must notify the Council at least 6 weeks in advance. This is to give the Council the opportunity to assess the contribution the tree makes to the character of the conservation area, and decide whether to make a Tree Preservation Order.
• Demolition: Demolition or substantial demolition of a building within a conservation area will usually require permission from the Council.

Endnotes
1 *Heritage and the economy,* 2015 Historic England.
2 *Energy efficiency and historic buildings: Application of Part L of the Building Regulations to historic and traditionally constructed buildings*, 2012 English Heritage.
3 Planning (Listed Buildings and Conservation Areas) (Scotland) Act 1997

1.2   Insurance

Post-completion insurance policies
Buildings insurance
Peter Morse

Insurance coverage basics include:

> **Who is protected**: the owner or occupier of a property, after practical completion.
> **What is covered**: all risks of physical loss/damage in respect of the reinstatement of the structure, out buildings, walls, gates, walkways including removal of debris, professional fees, etc.; rental income can also be included as can property owner's liability.
> **What is not covered**: maintenance issues, gradual deteriorations, defective workmanship, wear and tear, and building design defect.
> **Typical policy exclusions**: liability assumed under contract, financial loss, products supplied.
> **Duration of cover**: for the life of the structure, policies are usually arranged on an annually renewable basis; however, occasionally, long-term agreements can be made to secure a premium rate for a limited number of years.

Important points: financial institutions are increasingly taking an interest in the insurance arrangements and may ask to be included as a co-insured, which grants them the benefit from the policy in place. Most policies will contain a 'noted interest' clause, which automatically includes any parties with a financial interest.

Loss of rent derived from the property can also be covered under a property owner's policy. Rental income and/or costs for alternative accommodation is protected following an interruption or interference with the business as a result of damage caused by an insured peril. It is important to note that a loss of rent claim must follow a valid material damage claim (with a limited number of exceptions, e.g. denial of access).

All policies exclude cover for terrorism as standard; this can, however, be 'bought-back' for an additional premium. Therefore, if this cover is required, it is important to specifically ask for it to be included.

It is vital that the correct sums insured are provided when taking out a policy. All too often, the purchase price or constructions costs are used. These are usually too high (meaning that a higher premium is paid) or too low (which means that average will apply – a claim will be reduced by the percentage that the overall amount has been under-insured). The sum insured should be based upon an insurance valuation for the reinstatement of the entire structure following a total loss. This should include the building, any surrounding walls, pathways, costs for debris removal, and professional fees (VAT should also be added for residential buildings or where the owner is not VAT registered).

Insurers will usually look to restrict the insured perils on an unoccupied property (e.g. fire, lightning, explosion, earthquake, and aircraft), as it is perceived that this presents a greater risk. Cover on an all-risk basis can be arranged, but the premiums are likely to be much greater and there will be additional implications imposed on the owner. Usually they include regular inspections (which must be recorded), and removal of waste and combustible material on a regular basis. Most insurers would also expect an increased level of security including boarding windows and sealing letter boxes, and the disconnection and drainage of utilities.

Legal indemnity insurance
Rob Cooke

The ownership of property often involves the acceptance of risks that may be difficult to eliminate entirely but which nevertheless demand some measure of protection. This section gives a brief outline of the main insurance policy covers which can be obtained. It is, however, recommended that advice is sought from an insurance broker or adviser before cover is arranged.

Duration of cover: For legal indemnity policies, most insurers will provide cover in perpetuity and also permit transfer of the cover to successors in title.

Table 1.2

Indemnity name/type	Cover
Absence of easement	Policy protects against interference with the use of the property (the easement) due to a lack of legal grant for access and/or services to a property
Absent landlord	Policy protects against forfeiture of the lease of a property for non-compliance with covenants in this lease (and loss caused by an inability to enforce other parties' maintenance obligations contained in this lease – this is really maisonette indemnity)
Adverse possession	Policy protects against a challenge to occupation of land that forms part of the insured's property, but is not included in the documentary title to the property

Indemnity name/type	Cover
Breach of planning, or building regulations or listed buildings consent	Policy protects against enforcement action by a local authority in respect of lack of appropriate consents for alterations to or use of the insured's property
Breach of restrictive covenant	Policy protects against the financial consequences of a person successfully enforcing through legal proceedings a covenant burdening the insured's land
Building over sewer	Policy protects against increased costs of construction works under a build-over agreement with the local water authority and/or a sewer authority's requirement for demolition/alteration of the insured's property to allow access to the sewer
Chancel repair	Policy protects against a liability to contribute towards the repair of the chancel of a church
Contaminated land	Policy protects against financial costs should any historical contamination be discovered
Contingent buildings insurance	Policy protects against uncertain or inadequate insurance arrangements for the building of which the property forms part
Defective title	Policy protects against a third party attempting to: • enforce an estate right or interest adverse to or in derogation of the insured's title to the property; • prevent the insured's use of any right of way of easement necessary for the enjoyment of the property.
Enforcement of (known/unknown) adverse third party rights	Policy protects against the exercise of rights or easements over or under the property
Enlargement of lease indemnity	Policy protects against the exercise of historic rights, easements or other interests over the property contained or reserved in the original lease of the property
Flat/maisonette indemnity	Policy protects against uncertain or inadequate arrangements for maintenance of the building of which the property forms part

Indemnity name/type	Cover
Flying/creeping freehold (i.e. a freehold which overhangs or underlies another freehold)	Policy protects against uncertain or inadequate arrangements for maintenance of a flying/creeping freehold element of the property
Forfeiture of lease –bankruptcy/insolvency	Policy protects a mortgagee against forfeiture of their borrower's lease due to borrower's bankruptcy
Forfeiture of lease – breach of covenant	Policy protects a mortgagee against forfeiture of their borrower's lease due to breach of covenant
Freehold rent charge coverage	Policy protects against re-entry by a rent-charge holder due to non-payment of a rent-charge
Good leasehold	Policy protects against a challenge to the property's title and unknown covenants
Insolvency Act	Policy protects a mortgagee or successor in title to the transfer in the event the transfer of the property is set aside pursuant to the Insolvency Act
Judicial Review	Policy protects in the event that a planning permission is quashed
Mining/mineral rights indemnity	Policy protects against financial loss as a result of the future exercise of rights to extract mineral reserves underneath the property
Missing deeds	Policy protects in the event that ownership of the property is challenged and cannot be substantiated by reference to original title deeds
Possessory title indemnity (residential)	Policy protects against a challenge to the property's title
Search insurance	Policy protects against financial loss caused by a matter that would have been revealed to the owner of a property by a search conducted by the local authority relating to mining, drainage, village green
Town and village green	Policy protects against damages or compensation, costs of altering/demolishing, diminution in value, and abortive costs, where any third party (including any corporation) applies to the relevant statutory authority for the registration of the property or any part thereof as a town or village green after the policy date

2
Development and procurement

2.1 Contracts

Letters of intent
Suzanne Reeves

Parties involved in construction works often find that it is not commercially feasible to delay the commencement of works until a full contract has been agreed and signed. Instead, they choose to proceed on the basis of a letter of intent (or a pre-construction services agreement, as it is sometimes referred to) pending the completion of the full contract. 'Letters of intent' and 'pre-construction services agreements' are not legally defined terms, and in practice, such documents vary widely in their legal effect.

It is important to understand that, depending on its terms, a letter of intent may or may not create a binding contract between the parties. A good example of the contractual uncertainty that letters of intent can create is the case of *RTS Flexible Systems Ltd v Molkerei Alois Muller GmbH & Co KG* [2010] 1 WLR 753, which was finally decided by the Supreme Court in March 2010. In that case, the High Court, the Court of Appeal, and the Supreme Court all had different interpretations of whether (and on what terms) a contract had been agreed at the expiry of the letter of intent. The case shows that even where a letter of intent does give rise to a binding contract, the effect is sometimes not what was anticipated by at least one of the parties. It is therefore important that the letter of intent covers certain essential points and is carefully drafted so as to minimise the risk to both parties.

When would you use a letter of intent?

In *Cunningham v Collett* [2006] EWHC 1771 (TCC), Judge Coulson QC (as he then was) acknowledged that letters of intent were sometimes the best way of ensuring that works can start promptly, but only in appropriate circumstances, which he stated as follows:

- where contract workscope and price are either agreed, or there is a clear mechanism in place for such workscope and price to be agreed;
- where the terms and conditions are (or are very likely) to be agreed;
- where the start and finish dates and the programme are broadly agreed; and
- where there are good reasons for the works to be commenced in advance of the contract documents being finalised.

Judge Coulson did, however, comment that in his view, letters of intent are often used 'unthinkingly' in the construction industry and often simply to avoid difficult negotiations under the full contract. He observed that once a letter of intent was put in place and works started, there was a real risk that a full contract would never be put in place. This was exactly what happened in the recent case of *Ampleforth Abbey Trust v Turner & Townsend Project Management Ltd* [2012] EWHC 2137 (TCC) where numerous letters of intent were issued, none of which provided for liquidated damages. The employer successfully sued its project manager claiming that had a fully executed building contract been in place then the employer would have reached a more favourable outcome in its dispute with the contractor over delay to the project.

What provisions should go into a letter of intent from a developer's point of view?

Given the potential risk to the parties of entering into a letter of intent and a full contract materialising, the following provisions should be included as a bare minimum:

- Limiting the scope of the letter: By limiting the scope of the letter (either in time or money and ideally, both), the developer is not only encouraging the contractor to agree a full contract within a certain period of time or before a certain value of the works has been reached but also could be limiting its exposure should a full contract not be agreed or signed. The letter must, however, state clearly what happens when the limit is reached.

- In *RTS Flexible Systems* there was a dispute over whether the parties had intended a contract to come into existence after the termination of the letter of intent, and, if so, what terms had been agreed. The High Court held that a contract on limited terms had come into existence on the basis of the parties' conduct. The Court of Appeal rejected that, stating that no contract had been formed. The Supreme Court decided the matter finally by holding that a contract had come into existence but on wider terms than held by the High Court.

- A letter of intent should therefore state whether or not the contractor is to stop working once the limit has been reached. If he is to continue working the terms and the basis on which he is to be paid should be made clear, for example by reference to an agreed rate or by *quantum meruit* (an amount based on the value of the services provided calculated by reference to what is reasonable within the industry for such services).

- Identify the works to be carried out under the letter of intent: This can be used to limit the scope of the letter as mentioned above by authorising the contractor to carry out only a specified amount of work or place a specified number of orders.

- Insurance: If the contractor is to start works on site under the letter of intent, the letter should state the level of public and employer's liability insurance that must be taken out, as well as all risks insurance in respect of the works themselves. If the contractor is to carry out design services under the letter then the required level of professional indemnity insurance should be stated.

- CDM Regulations requirements: There should be a contractual requirement for the contractor to comply with the requirements of the CDM Regulations and, if appropriate, act as Principal Contractor.

- Payment provisions: These must be clearly stated or incorporated in order to avoid default payment provisions of the Scheme of Construction Contracts applying.

- Warranties: Depending on the type of works to be carried out under the letter of intent, third parties with an interest in the development may require a collateral warranty from the contractor and its design subcontractors. If so, the letter must provide for this in its terms.

- Copyright: If the contractor is carrying out any design services under the letter of intent, then it should provide for a licence to be given to the developer allowing it to use any design documents produced by the contractor in connection with the development. This is important, for example, in circumstances where the contractor's involvement in the development is terminated but the developer still needs the contractor's design documents.

- Termination: The letter of intent must provide for the developer to terminate the contractor's engagement under the letter should it so wish. The letter should also make it clear that the developer is under no obligation to enter into a full contract with the contractor once the letter of intent has come to an end.

Summary

- Letters of intent should be treated with caution. As Lord Clarke said in *RTS Flexible Systems*, "The moral of the story is agree first and start work later".
- Parties should therefore give careful thought as to whether a letter of intent is appropriate, having regard to the circumstances suggested by Judge Coulson.
- A letter of intent, depending on its terms and the conduct of the parties, may or may not give rise to a binding contract.
- Therefore, if a letter of intent is deemed to be appropriate, it should be well drafted and precise, so that the parties are not left in a position of uncertainty prior to the execution of the full contract.
- If the letter is not drafted carefully, covering the main areas of risk, the parties may be unnecessarily exposed, both during the course of the works and when the letter comes to an end.

2.2   Contract management

Employer's agent
Tim French

Clients routinely wish to procure projects in the pursuit of certainty of cost. The JCT Design and Build contract, which is in frequent use, transfers most of the 'risk' in any development to the contractor, while allowing the contractor greater flexibility to deliver the product. As an employer's agent, it is essential that certain pre-contract and post-contract services be provided.

Pre- and post-contract services suggested to provide include (but are not limited to) the following:

Pre-contract services

- Define the responsibilities of the employer, employer's agent, and contractor.
- Appraise and quantify the risks.
- Formulate the employer's brief and identify specific requirements.
- Assess the contractor's proposals and ensure compliance with the employer's requirements.
- Undertake a design audit of the contractor's proposals for compliance with the employer's requirements.
- Evaluate the offer, the contract sum analysis and stage payments, and assess value for money.

Post-contract services

- Generally comply with the duties of the employer's agent under the building contract.
- Set up quality control procedure and report on works carried out on site. This may require the appointment of monitor design consultants on larger contracts.
- Undertake site visits and chair meetings.
- Implement changes to the employer's requirements only on written approval of the client.
- Agree stage payments and recommendations for payments.
- Prepare monthly project control statements and cash flow forecasts to client.
- Advise on practical completion, preparation of snagging schedules, and component literature.
- Make sure operation and maintenance manuals and/or purchaser packs are prepared and approved by appropriate consultation.

General exclusions

- Checking and verifying contractor's design in terms of adequacy and efficiency.
- Checking and verifying contractor's design in terms of fitness for purpose.

In undertaking duties as the employer's agent, it is important to recognise the contractor's freedom to design while respecting the client's brief and auditing the quality of the end product. Make sure the employer's requirements are sufficiently tightly drawn up as to ensure the contractor delivers the project to meet the client's expectations.

The quantity surveyor's role
Tim French

The quantity surveyor (QS) is tasked with controlling construction cost by accurate measurement of the works and the application of expert knowledge of costs of labour, materials, and plant. A good QS can aid in an understanding of the implications of design decisions at an early stage, which ensures that good value is achieved and that clients receive accurate advice.

A skilled QS needs:

- an ability to predict future costs from limited information at an early stage in a project;
- an ability to manage the procurement process to ensure that predictions of cost, time, and quality are delivered;
- an ability to accurately value works carried out under a building contract;
- an awareness of risk with a capability to assess and manage risk;
- an ability to demonstrate value for money assisted by effective value engineering;

- the ability to communicate effectively with the client, design team, and contractor throughout the life of a project;
- an understanding of how the client's business impacts the project; and
- an awareness of financial incentives and opportunities available to the client (such as VAT relief and capital allowances).

The QS will be responsible for preparing the tender documents, reviewing and finalising tenders, and preparing the contract documents.

Most quantity surveyors will be employed either by construction consultants, main contractors, or subcontractors. They may also act as expert witnesses under construction disputes. The roles vary considerably; the consultant role is often that of independent certifier, whereas when acting for main contractors and subcontractors, quantity surveyors are more likely to be representing their employer's commercial interests.

Current trends in the industry continue for cost reductions, with pressure coming from major purchasers – the importance of value for money remains the prime objective for most projects. Fixed out–turn costs are a necessity.

This is against a backdrop of an anticipated general reduction in workload in all sectors. This scenario exerts additional pressure on the robustness of procurement information produced by the quantity surveyor.

The QS expertise is often useful to clients for the management and administration of projects where a detailed knowledge of building contracts is useful to protect the interests of the parties.

2.3   Cost management

Reinstatement valuations for insurance purposes
Tim French

Generally it is considered prudent to insure a property for the value of complete reinstatement following total destruction (including partial destruction that necessitates demolition and rebuild). The calculation of construction costs can take different forms depending on the client brief and the information available. More complex or bespoke properties may benefit from detailed cost planning, while for simple structures a cost per m^2 is sufficient.

The 'day one' value is effectively the cost of the reinstated building as completed and the final account settled. This is known as the 'declared value'.

By contrast, the 'insured value' is the declared value plus allowances for inflation during the policy term and rebuild period. This is normally decided following consultation with the insurance broker.

When preparing a reinstatement valuation, the following points should be considered:

- Include demolition cost.
- Rebuilding costs to be valued at 'day one' rates – using rates applicable at the time of preparation of the valuation; 'day one cost' can also be referred to as 'declared value'; alternatively 'insured value' may be stated, which would include an allowance for inflation.
- Include for any special features that are to be included to enable replacement of the property on a like-for-like

basis, for example, facade treatments (such as stonework embellishments) or internal features (such as specialist decorations).

- Ensure that the scope of fixtures, fittings, and installations is clearly defined, particularly in tenanted properties.
- Include professional fees at an appropriate level, depending on the complexity and location of the property.
- Account for geographical location factors either in rebuilding rates or separately by reference to recognised indices.
- Include local authority planning and building regulation fees, as they are unavoidable expenses.
- The cost of rebuild should include for the impact of latest legislation, such as renewable energy installations. If the valuation is being projected to cover the period of the policy (as opposed to 'day one' valuation) then account should be made for the:
  - period of the policy;
  - design period;
  - planning period;
  - construction period; and
  - void (letting) period (if required by the owner).

This may result in projecting costs for anything up to and beyond a 3-year period. A contingency should be included.

Reinstatement valuations for listed buildings

Listed buildings should be reinstated to the same design, quality, style, and workmanship, and in the same material, but in accordance where necessary, with current legislation. Historic buildings, therefore, tend to be more expensive to reproduce than modern equivalents, and hence more expensive to insure.

The building should be surveyed to establish the scope and specification of both the external and internal construction and finishes; this may also include archival research. A photographic report should be undertaken as a record of the survey and findings.

A design statement should be prepared detailing exactly how, in the event of loss, the building would be reconstructed. This should pick up the reconstruction of the shell, as well as all the historic features and points of historic interest and merit. A competent person, who can interpret the likely requirements of a conservation officer, should undertake the design statement.

A detailed estimate or cost plan should then be prepared to pick up the level of detail necessary to accurately price the scope contained within the design statement. The estimate should allow for appropriate skilled tradesmen and materials commensurate to the existing building.

It is unlikely that a reinsurance valuation based on unit rates multiplied by floor area will provide a reliable valuation.

2.4   Building and construction

Dispute resolution
Suzanne Reeves

Disputants and their advisors have a variety of dispute resolution mechanisms that they can select to resolve their disputes. The principal dispute resolution procedures are:

- Adjudication
- Arbitration
- Early neutral evaluation (ENE)
- Independent expert

- Litigation
- Pre-Action Protocol for Construction and Engineering Disputes (the Protocol)
- Mediation
- Dispute review boards

Adjudication

This process is enshrined in the Housing Grants Construction and Regeneration Act 1996 (as amended by Part 8 of the *Local Democracy, Economic Development and Construction Act* 2009). A wide variety of the disputes arising under construction contracts can be referred to adjudication. From 1 October 2011 in England and Wales, disputes regarding oral contract can also be referred to adjudication. The Act does not deal with disputes with residential occupiers unless the parties agree that adjudication will apply. Adjudication is popular not least because an adjudication award has to be made within 28 days of the case being referred to an adjudicator unless the parties agree otherwise. The decision of an adjudicator is binding pending agreement of the parties or the decision of a court or arbitrator.

Arbitration

Arbitration is based on the contractual provisions agreed by the parties. Procedures under the Arbitration Act 1996 have given the arbitrator wide powers to resolve disputes without unnecessary cost or delay, and in a fair manner without undue interference from the courts. The right of appeal is limited.

Early neutral evaluation (ENE)

Technology and Construction Court (TCC) judges are prepared to arrange a short hearing of a case or alternatively by paper review only, on specific issues usually on a without prejudice basis (but this can be waived by agreement) and give preliminary views on the merits, as an aid to settlement discussions between the parties. If a judge determines a particular issue by ENE the parties are free to agree whether or not they will be bound by it. If the ENE does not result in settlement, the case can proceed to trial but will be heard by another judge with no knowledge of the outcome of the ENE.

Independent expert

This procedure forms a valuable means for the speedy resolution of technical disputes, and is generally straightforward and flexible. Issues in dispute are referred to an expert to decide using his or her own professional expertise or judgement. It has been successfully used for many years in rent-review matters, but has much wider application to technical disputes. If the parties agree to be bound by the expert's decision the right of appeal exists only in limited circumstances only.

Litigation

The Civil Procedure Rules 1998 have led to more efficient running of cases both in terms of cost and time. The Technology and Construction Court in the High Court deals with construction disputes, and has considerable experience doing so. Before court proceedings are commenced, the parties should comply with the Pre-Action Protocol for Construction and Engineering Disputes.

Pre-Action Protocol for Construction and Engineering Disputes (the Protocol)

This applies to all construction and engineering disputes (including professional negligence claims against engineers, architects, and quantity surveyors). It is not intended to apply to debt claims where it is not disputed that the money is owed and where the claimant follows a statutory or other formal pre-action procedure.

The objective of the Protocol is to encourage the exchange of information about the prospective claim so that parties can narrow the issues between them and explore settlement options in an attempt to avoid litigation. Where litigation cannot be avoided, the Protocol aims to support the efficient management of proceedings.

The Protocol should be read with Section 2 of the Technology and Construction Court Guide and The Practice Direction on Pre-action Conduct and Protocols.

The Protocol sets out a number of steps that should be followed before proceedings commence, including attention to the letter of claim, the defendant's response, and the pre-action meeting.

Letter of claim

The claimant serves a letter of claim to each proposed defendant detailing the following information:

The claimant's full name and address
The full name and address of each proposed defendant
A summary of the facts of the claim
The basis on which each claim is made, including the principal contractual terms and any statutory provisions relied on
The nature of the relief claimed
If the defendant(s) previously rejected the claim, the claimant's grounds of belief as to why the claim was wrongly rejected
The names of any experts instructed by the claimants and the issues that will be addressed

The defendant's response

The defendant must acknowledge receipt of the letter of claim in writing within 14 days. If the defendant fails to do this, the claimant is entitled to commence proceedings without further compliance with the Protocol.

Within 28 days from receipt of the letter of claim, the defendant is to send a letter of response to the claimant containing the following information:

- the facts set out in the letter of claim, which are agreed or not agreed, and if not agreed, the basis of the disagreement;
- which claims are accepted and which are rejected, and the basis of the rejection;
- if a claim is accepted, whether the damages, sums, or extension of time are accepted or rejected, and if rejected, the basis for the rejection;
- if contributory negligence is alleged, a summary of the facts relied upon;
- whether the defendant intends to make a counter-claim, and if so, giving the information detailed in the above points; and
- the names of any experts instructed and the issues that will be addressed.

The time for serving the response can be extended by agreement between the parties by up to 3 months. If the defendant has made a counter-claim then the claimant should respond within the same period that the defendant was given to respond to the letter of claim.

Pre-action meeting

The Protocol requires the parties to meet on a without-prejudice basis at least once. This meeting should happen within 28 days of the claimant receiving the defendant's letter of response, or after the defendant has received the response to the counter-claim.

The aim of the meeting is to try and resolve the dispute without litigation. The parties should seek to agree the main issues in the case and identify the cause of the disagreement in respect of each issue. If the parties cannot settle the dispute at the meeting, they should agree the steps that should be taken in the litigation to ensure it is conducted in accordance with the overriding objective in Rule 1.1 of the Civil Procedure Rules.

Paragraph 4.6 of the *Practice Direction – Pre Action Conduct* states that if a party fails to comply with the Protocol, the court may do the following:

- Adjourn proceedings until the steps set out in the Protocol have been taken
- Order that the party at fault pay part or all of the costs of the other party
- Order that the party at fault pay those costs on an indemnity basis
- Award interest to a successful party that did not comply at a lower rate than would otherwise have been awarded
- Award interest against an unsuccessful party that did not comply at a higher rate than would otherwise have been awarded (up to 10 per cent above the base rate)

However, it is important to note that in the recent case of *Higginson Securities (Developments) Ltd and another v Kenneth Hodson (2012) EWHC 1052 (TCC)*, the defendant was not entitled to a stay following the claimant's noncompliance with the protocol. The TCC interpreted Paragraph 5.1 of the Protocol that states "the parties should normally meet" to mean a meeting should take place unless there was a reasonably good reason for it not to.

A TCC working party is currently reviewing the Protocol following the recommendations set out in Lord Jackson's report: *Review of civil litigation cost.* The working party is considering the future role of the Protocol, in particular whether it should be retained, abolished, or amended to be voluntary rather than compulsory.

Mediation

Mediation is a voluntary and non-binding procedure. It is a private process in which an independent neutral person helps the parties reach a negotiated settlement. The Mediator usually does not make a determination on the dispute but may do so if the parties agree and he or she considers that it will assist in reaching a settlement.

Dispute review boards

Dispute review boards (DRBs) are used on larger projects. A panel is appointed at the start of the project and visits the site a number of times per year to deal with disputes by providing interim binding decisions (similar to an adjudicator). The decisions can be challenged via litigation or arbitration.

Part II of the Housing Grants, Construction and Regeneration Act 1996
Suzanne Reeves

The following aims to set out a brief outline of the issues which need to be considered when determining whether contractual terms are compliant with Part II of the Housing Grants, Construction and Regeneration Act 1996 (the Act) as amended by the Local Democracy, Economic Development and Construction Act 2009 (LDEDCA) or whether certain provisions will be incorporated into the contract by the Scheme for Construction Contracts.

Parties to a construction contract are free to negotiate and agree the terms and conditions under which the works and services are to be carried out. However, there are times where a contract fails to comply with the minimum requirements relating to adjudication and payment laid down by the Act. Consequently, certain provisions will automatically be incorporated into the contract by the Scheme for Construction Contracts. The original Act provisions will apply to construction contracts entered into before 1 October 2011. Contracts entered into on or after 1 October 2011 will be subject to the Act as updated by the provisions of LDEDCA. The principal amendments made by LDEDCA include:

- Parties who have an oral or partly oral contract will be able to rely on the provisions of the Act.
- The payment and withholding notices provisions are overhauled. For example, either party (rather than only the payer) can issue a payment notice stating how much is due, and payment notices must be given even if the amount due is zero. If no notice to pay less is given then the notified sum becomes due.
- There is a provision in the Act, as amended by LDEDCA, dealing with pre-agreeing the allocation of costs of future adjudications in the contract (known in the industry as 'Tolent' clauses). The intention is that clauses which provide that the party bringing an adjudication claim will be liable to pay all costs of the adjudication will be outlawed unless agreed by both parties at the time of the adjudication. Arguably, however, the relevant legislation wording is ambiguous and the courts may need to interpret the legislation before it can be said with certainty that such clauses are illegal.
- 'Pay when certified' clauses will be prohibited in most contracts. For example, the main contractor cannot make payment to its subcontractors conditional on its own payment by the employer being certified. This also applies at all other levels of the construction chain, including consultants.

- The suspending party can claim the costs from the exercise of the right to suspend and can claim an extension of time to complete its work for the delays resulting from the exercise of the right.
- Contracts will need to include a right for the adjudicator to correct typographical or clerical errors (such as miscalculations) in his or her decision (the courts had previously confirmed this right, the 'slip rule').

When does the Act apply? (S104–107)

The Act applies to:

- contracts entered into on or after 1 May 1998;
- contracts entered into on or after 1 October 2011 are subject to the Act but as amended by LDEDCA;
- contracts in writing – it is sufficient if the contract is evidenced in writing, and for contracts entered into on or after 1 October 2011, the Act (as amended by LDEDCA) will apply to oral contracts;
- contracts for construction operations which include the construction, alteration, repair, maintenance, decoration, demolition, and installation in buildings forming or to form part of the land, and also architectural, design, surveying, or engineering advice; and
- the carrying out of construction operations in England, Wales, or Scotland, whatever the applicable law of the contract.

The Act does not apply to:

- contracts with residential occupiers for work on their property where they intend to occupy the property as their residence;
- certain mining, drilling, and extraction operations;
- installation or demolition of plant or machinery or steelwork to support or provide access to plant or machinery

on a site where the primary activity is nuclear processing, power generation, water or effluent treatment, or the production processing of chemicals, pharmaceuticals, oil, gas, steel, food, or drink;
- manufacture and delivery of materials not involving installation (supply only); and
- artistic works.

Letters of intent may be subject to the Act where they are sufficient to amount to a legally binding contract in their own right.

Adjudication (S108)

The contract must:

- allow either party to refer a dispute to adjudication at any time;
- provide for the appointment of an adjudicator within 7 days;
- require the adjudicator to reach a decision within 28 days after the dispute has been referred to the adjudicator (or a longer period if agreed by the parties);
- allow the adjudicator and the party who referred the dispute to the adjudicator to extend the period for the decision by up to 14 days;
- impose a duty on the adjudicator to act impartially;
- enable the adjudicator to take the initiative in ascertaining the facts and the law surrounding the dispute;
- provide for the decision of the adjudicator to be binding on the parties until the dispute is taken to arbitration or the courts;
- provide that the adjudicator is not liable for anything he or she does unless he or she acts in bad faith; and
- for contracts entered into on or after 1 October 2011 (where the Act as amended by LDEDCA applies), include a provision permitting the adjudicator to correct his or her

decision to remove a clerical or typographical error arising by accident or omission.

If the contract does not comply with the adjudication provisions of the Act in full, all the adjudication provisions of the contract will be set aside and the adjudication procedures under the Scheme for Construction Contracts will apply. The procedures under the Scheme cover the points listed above and introduce time limits.

These provisions must be in writing; therefore if the contract is entirely oral (and after 1 October 2011 the Act, as amended by LDEDCA, applies to oral contracts) the adjudication procedures under the Scheme for Construction Contracts automatically apply.

Adjudication costs (S108A)

LDEDCA has introduced a new section 108A to the Act. The intention of this provision was to ensure that parties could only agree the allocation of the costs of an adjudication after the date of the relevant adjudication notice. This was intended to preclude what is known in the industry as a 'Tolent clause' (a clause which requires the party bringing the adjudication to pay the costs). There is concern, however, that the wording of the legislation is ambiguous and that parties could still use Tolent clauses in their contracts. The courts may have to interpret this provision to determine whether or not Tolent clauses are illegal.

Payment

The Act and the Scheme for Construction Contracts provide a 'menu' of payment provisions covering:

* payment by instalments;
* final payment;

- withholding payment; and
- conditional payment.

If a contract fails to comply with any one of the provisions from the 'menu', the relevant provisions from the Scheme will apply. The remainder of the contractual provisions that do comply with the Act will remain intact. It is therefore possible to end up with a contract where the payment provisions are a mixture of express terms agreed between the parties and implied terms from the Scheme.

Payment by instalments (S109–110/110B)

A party to a construction contract is entitled to payment by instalments, stage payments, or other periodic payments unless:

- the contract specifies that the duration of the work is less than 45 days; or
- the parties agree that the work is estimated to take less than 45 days.

Where the work falls within the 45-day limit, the right to instalment payments is excluded, but all other payment provisions (notice of withholding payment, set-off, and the like) will apply, as will the adjudication provisions outlined above.

Where a contract falls below the 45-day limit, payment of the contract price falls due 30 days after completion of the work (or 30 days after the contractor's claim if later) and payment must be made within 17 days.

In all other cases, the parties can agree between themselves:

- the amounts of each payment;
- the intervals between each payment;
- the date each payment becomes due; and
- the final date by which each payment must be made.

If the contract does not contain a clear mechanism for determining each of these four elements they will be determined by the Scheme for Construction Contracts, namely:

- The amount of each payment will be based on the value of the work and other costs to which the contractor is entitled during the payment interval.
- There will be 28-day payment cycles.
- Payment is due 7 days after each 28-day period (or 7 days after the contractor's claim for payment if later).
- The final date for each payment is 24 days after each 28-day period (or 24 days after the contractor's claim for payment if later).

The contract must provide for the paying party to give notice within 5 days of the date on which each instalment becomes due, specifying the amount proposed to be paid and the basis on which it is calculated (for contracts to which the Act as amended by LDEDCA applies it is immaterial whether the amount due is zero). Any attempt in the contract to vary or exclude this requirement will be ineffective and this provision will be implied by the Scheme for Construction Contracts.

In contracts entered into on or after 1 October 2011, if the paying party (or another party specified in the contract as being responsible for giving the payment notice) fails to give such notice, then the payee may give the notice specifying how much the payee considers to be due and the basis on which that amount is calculated.

Final payment

The contract must contain a clear mechanism for determining when the final payment due under the contract becomes payable and the final date by which that payment must be made.

The parties are free to agree the dates or periods within which the final payment is due and is payable but if there is no such mechanism, in accordance with the Scheme, the final payment:

- is due 30 days after completion of the work (or 30 days after the contractor's claim for payment if later); and
- the final date for making the final payment is 47 days after completion of the work (or 47 days after the contractor's claim for payment if later).

Withholding payment (S111)

No payment can be withheld unless a 'notice of intention to withhold payment' has been given for contracts entered into before 1 October 2011 where the original Act applies, specifying the amount to be withheld and the grounds for withholding payment. For contracts entered into on or after 1 October 2011 which are subject to the Act as amended by LDEDCA, the payer (or other specified person) must serve a notice to 'pay less', specifying the amount the payer considers due and the basis on which that sum is calculated (i.e. more information is required for a pay-less notice than was required under the original withholding notice regime).

The notice must be given before the final date for payment.

The contract can specify how long before the final date for payment the notice must be given (even if it is just one day) but if this notice period is not specified, the Scheme applies and at least 7 days notice must be given.

Conditional payment (S113)

Any provision in a contract which makes payment conditional upon the paying party receiving payment from someone else (and also for contracts entered into on or after 1 October 2011, payment being certified as due under a superior contract) is ineffective and the payment provisions of the Scheme outlined above will apply.

The only exception is where the contract provides that payment may be withheld if the reason for non-payment is the insolvency of someone else in the payment chain.

Suspending performance (S112)

If any payment is not received by the final date for payment and a notice of withholding payment or a 'pay less' notice has not been served, the contractor may suspend work after giving 7 days notice of its intention. The right to suspend performance ceases when the relevant payment is received. The period in which to complete the works is automatically extended by the number of days of the suspension.

This brief summary of Part II of the Housing Grants, Construction and Regeneration Act 1996 is not intended to be a detailed explanation of the provisions, and we recommend that legal advice is sought on any specific issues.

Insurance policies pre-/during construction
Rob Cooke

This section gives a brief outline of the main policy covers which can be obtained. However, it is recommended that advice is sought from an insurance broker or adviser before cover is arranged.

Professional indemnity insurance (PII)

Who is protected: architects, engineers, surveyors, project managers, designers, and other professionals.

What is covered: losses resulting from professional negligence, errors and/or omissions which cause financial loss to a third party.

What is not covered: material damage, theft, personal injury, and damage to third-party property.

Typical policy exclusions: work carried out prior to the inception of the policy, insured v insured claims (i.e. where the board of a company will seek damages against an employee for their professional negligence which causes a loss to the company), and insolvency.

Duration of cover: 1 year – annually renewable contracts are usually arranged. Policies are placed in to run-off should the professional cease trading.

Important points: professional indemnity is usually underwritten on a claim-made basis. This means that it is the policy at the time a claim is made which will respond and not the policy in place at the time of the initial error or omission. It is also essential to ensure that the cover provided under this policy extends to include collateral warranties.

Depending on the contract, cover should usually be maintained for at least 6 (usually 12) years after the completion of the contract.

Collateral warranties are frequently executed under seal which increases the limitation period from 6 to 12 years. Contractors may also find themselves responsible for their specialist subcontractors (e.g. plumbers, electricians, roofers, scaffolders, and piling contractors).

Most professional indemnity insurers accept that collateral warranties are used quite regularly and some have taken positive steps to accept certain warranties and to give helpful advice.

Employer's liability (EL)

Who is protected: building/engineering contractors, including any bona fide subcontractors.

What is covered: death, injury, or illness to employees as a result of the employer's negligence.
What is not covered: people on sick leave or action brought at an Employment Tribunal.
Typical policy exclusions: none, however restrictions as to the type of work that is covered may apply.
Duration of cover: 1 year – this type of policy is usually arranged on an annually renewable basis.
Important points: EL is a compulsory insurance under statute.

The minimum legal requirement is a limit of indemnity of £5m, however most insurers provide £10m as a standard. Even a limited company with only one employee is required to have employer's liability insurance.

Cover of employees is extended to include, volunteers, labour-only subcontractors, working principals, apprentices, and young people on work experience.

Under new legislation it is now a requirement of EL policies to be registered with the Employers' Liability Tracing Office (ELTO, www.elto.org.uk) in order to be able to identify the correct insurer in the event of any claims being made a few years down the line. Additionally due to the ELTO regulations, Employee Reference Numbers (ERN) must be supplied to insurers at the inception or renewal of the policy, this includes the ERN of all subsidiary companies.

Public/products liability insurance

Who is protected: building/engineering contractors.
What is covered: third-party property damage or injury, resulting from negligent acts or omissions or caused by the supply of faulty or defective products.
What is not covered: material damage, theft, and pure financial loss suffered by third parties.
Typical policy exclusions: injury to employees, work involving explosives, defective workmanship.

Duration of cover: at all times during the contract. Annually renewable contracts are usually arranged.

Important points: products liability is usually provided on an aggregate basis and not for each and every claim. This means that the limit of indemnity provided will be for all claims occurring during the policy period and not for each claim. It is therefore important to set the correct limit from the outset in order to ensure there is adequate cover for any potential claims.

Most policies will also contain a bona fide subcontractor's clause (BFSC); this means that any BFSC must carry the same limit of indemnity as the main contractor in order of the policy to respond.

Cover will automatically extend to include acts of labour-only subcontractors and indemnity to principals, together with any costs relating to legal fees, expenses, and hospital treatment, including ambulance costs that the NHS may claim.

Non-negligent liability insurance

Who is protected: building/engineering contractors and their employers in joint names.

What is covered: damage to third party property as a result of collapse, subsidence, heave vibration, weakening or removal of support, or lowering of ground water in the normal course of the works being carried out and for which no party can be found to be negligent.

What is not covered: any damage or injury for which negligence can be established, damage to the works or materials.

Typical policy exclusions: defective design or workmanship, inevitable damage.

Duration of cover: at all times during the contract, cover arranged on a project specific basis.

Important points: this type of cover is sometimes referred to by the reference in the JCT standard contract as 21.2.1 or 6.5.1 cover.

Because of the nature of the work, it is possible that adjoining properties to the contract site may be damaged following the activities of the contractor. However, the contractor may not have carried out the work negligently. The contractor's public liability insurance policy deals with allegations of negligence only and without evidence from the third party that the contractor had been negligent, their claim may well fail.

However, the employer (developer) may still be liable, as it will be seen as the party that brought the contractor to site (*Gold v Patman & Fotheringham* [1958]), hence the need for this more specific insurance cover.

It is the employer's responsibility to ensure cover is taken out but it is usually easier for the contractors to arrange this as an extension to their public liability insurance and charge the premium back to the employer. This prevents any disputes between the public liability and non-negligent liability insurers in the event of a claim.

This cover is usually arranged on a project-by-project basis. However, a limited number of insurers will offer an annual policy (usually based on declarations from the contractor).

Contract works/contractors all risk/erection all-risks policies

Who is protected: building/engineering contractors and their employers in joint names.
What is covered: 'all risks' of loss to the works and/or contractors' plant, materials, and the existing structure being worked on (if applicable).

What is not covered: injury to employees or third parties, third party property damage, defective workmanship, or defective materials.

Typical policy exclusions: disappearance and shortages, damage occurring after practical completion and after the maintenance period.

Duration of cover: 1 year – renewable annually. At all times during the contract and any maintenance period after practical completion.

Important points: attention should be paid to the selected clauses within the contract. Under JCT (or similar) contacts, several options are available in terms of who becomes responsible for insuring the works and existing structures (sometimes referred to as Clause 22 or 6.7):

Option A: for new builds, the contractor is responsible for arranging an all risks policy in joint names to cover the works and professional fees.

Option B: for new builds, the employer is responsible for arranging an all-risks policy in joint names to cover the works and professional fees.

Option C: for refurbishment/renovation, the employer is responsible for arranging an all-risks policy in joint names to cover the works, and professional fees and a specified perils policy, again in joint names for the existing buildings.

Joint names policies contain subrogation waivers, which means that the insurer cannot recover its costs from the party responsible for the damage caused.

When selecting the contract sum insured, various factors should be taken into consideration, not just the value of the materials used. Cover should be arranged to include professional fees, costs of removing debris, and VAT.

Contractors will be responsible for insuring their own or hired-in plant against theft or other types of loss. It should also be arranged that contractors are responsible for insuring against their own risks of employer's and public liability, and that all BFSC carry adequate levels of insurance.

Cover can be arranged on a project-specific basis, which is usually the case if a joint-names policy is required or for a larger contract that would normally fall outside of the contractor's usual activities or would be extended beyond the policy period and therefore would not be covered under an annual contractors' all-risks policy.

Environmental impairment liability insurance

Who is protected: property owners and developers; environmental engineers and consultants; plastics manufacturers; businesses with own oil tanks or diesel supplies; businesses involving hazardous chemicals; general manufacturing.

What is covered: environmental damage and/or prevention and remediation whether from gradual or sudden and accidental events. It includes covers such as first party (own site) clean-up costs (resulting from gradual and sudden and accidental pollution), third party (off site) clean-up costs (resulting from gradual and sudden and accidental pollution), third party nuisance claims, transportation, biodiversity damage, (Environmental Damage Regulations) coverage, business interruption, mitigation expense, and defence costs.

What is not covered: known conditions, microbial matter, abandoned property, asbestos and lead (but covers clean up), intended or expected loss, identified underground storage tanks (unless specified with insurers), intentional non-compliance with legislation, insured vs insured, prior knowledge, and non-disclosure.

Typical policy exclusions: employer's liability, contractual liability, fines, and penalties.

How long should cover be maintained: annually renewable policies are available for operational business risks and for property owners. Special requirements policies with periods of 3, 5 and 10 years are the norm.

Important points: under Environmental Damage Regulations, operators (property owners, business operators, etc.), are strictly liable whether or not the operator intended to cause damage or was negligent. The basic principle is that the polluter pays. The Environment Agency has powers to rectify pollution damage and then recover the costs from the polluter.

Conventional public liability policies carry a standard exclusion in respect of gradual pollution and contamination; the policy only indemnifies against pollution claims as a result of sudden and unforeseen events. Gradual pollution losses are excluded so a specific environmental impairment liability policy is needed.

The Construction (Design and Management) Regulations 2015
Paul Gibbison and Rebecca Fairclough

The Construction (Design and Management) Regulations 2015 (CDM 2015) came into force on 6 April 2015 to replace the previous CDM Regulations 2007 (CDM 2007) and apply to all building and construction projects, regardless of the size, duration, and nature of the work.

The Construction (Design and Management) Regulations are the main set of regulations for managing the health, safety, and welfare of construction projects. They provide a statutory framework for managing health and safety during the construction, repair, maintenance, and demolition of civil engineering and construction work; with legal duties for everyone involved in the project from design through construction and beyond, including the maintenance, use, and demolition of the building or structure.

Key aims

The key aims of the CDM Regulations are to:

- sensibly plan the work so the risks involved are managed from start to finish;
- have the right people for the right job at the right time;
- cooperate and coordinate with the design and construction project team;
- have the right information about the risks and how they are being managed;
- communicate this information effectively to those who need to know; and
- consult and engage with workers about the risks and how they are being managed.

The main changes

The main changes from CDM 2007 to CDM 2015 are:

1. Simplification of the regulations. The regulations are clearer shorter than the 2007 version.
2. Removal of the Approved Code of Practice. This has been replaced with a new guidance document (HSE L153 Publication).
3. The CDM Coordinator role has been replaced and its duties have now been transferred onto the new Principal Designer role and the Client.
4. Explicit competence requirements have been removed (old Appendix 4), and replaced with "skills, knowledge and experience".
5. The new regulations have introduced the domestic market which was exempt from previous regulations.
6. The threshold for F10 notification* has changed to:

* The F10 notification is an electronic document to be completed and submitted on the HSE website when works exceed the threshold identified above.

Table 2.1 Duty holders – Overview of differences

Duty Holder	CDM 2007	CDM 2015
Client	• Check competence • Ensure management arrangements • Ensure sufficient time & resources • Provide Pre-construction information (PCI) • Appoint a CDM-Coordinator (CDMC) • Appoint a principal contractor (PC) • Ensure suitable construction phase plan (CPP) is in place prior to commencement • Adequate welfare facilities on site • Provide information relating to the health and safety file to the CDMC • Retain and provide access to the health and safety file	• Check competence • Ensure management arrangements • Ensure sufficient time & resources • Provide Pre-Construction information (PCI) • Appoint a Principal Designer (PD) where there is more than one Contractor • Appoint a Principal Contractor (PC) where there is more than one Contractor • Ensure Constuction Phase Plan (CPP) is drawn up prior to commencement • Adequate welfare facilities on site • Check that the H&S File has been prepared by the PD (or PC) • Retain and provide access to the health and safety file • F10 Notification

Principal Designer (PD)	• N/A	Required where there is more than one contractor
		• Plan, manage and monitor the Pre-construction phase including:
		  – Take account of the general principles of prevention
		  – Ensuring designers carry out their duties
		  – Liaise / cooperate with the PC and others
		• Assist the Client in preparation of the PCI
		• Health and safety file

Role		
Designer	<ul><li>Eliminate hazards and reduce risks during design</li><li>Provide information about remaining risks</li><li>Check that the client is aware of their duties and that a CDM Co-ordinator has been appointed.</li><li>Provide information needed for the health and safety file</li></ul>	<ul><li>Eliminate hazards and reduce risks during design</li><li>Provide information about remaining risks</li><li>Check that the Client is aware of their duties</li><li>Provide information needed for the health and safety file</li></ul>
Principal Contractor	(Required on notifiable jobs)<ul><li>Plan and monitor</li><li>Prepare, issue, implement a written plan</li><li>Share the plan</li><li>Provide welfare</li><li>Check competence</li><li>Induct and train</li><li>Consult</li><li>Liaise with CDM-C</li><li>Secure the site</li></ul>	(Required where there is more than one contractor)<ul><li>Plan, manage and monitor</li><li>Prepare, issue, implement a CPP</li><li>Share the plan</li><li>Provide welfare</li><li>Check competence</li><li>Induct and train</li><li>Consult with the workforce</li><li>Liaise with the PD</li><li>Secure the site</li><li>Provide Health & Safety File if PD appointment has ended before project completion</li></ul>
Contractor	<ul><li>Plan and monitor</li><li>Check competence</li><li>Train</li><li>Provide information</li><li>Ensure welfare</li><li>Cooperate with the PC</li><li>Provide information for HSF</li><li>Inform PC of accidents</li><li>Information PC of issues with the CPP</li></ul>	<ul><li>Plan, manage and monitor</li><li>Comply with the PC and PD and the relevant parts of the CPP (where there is more than one Contractor)</li><li>If only one Contractor, take account the general principles of prevention and draw up a CPP</li><li>Check competence</li><li>Train, instruct and provide information</li><li>Secure the site</li><li>Provide welfare facilities</li></ul>

a 30 days on site <u>AND</u> more than 20 workers working simultaneously at any point in the project; or

b Exceeds 500 person days (e,g, 50 workers working simultaneously for 10 days = 500 person days).

c "Construction" definition has expanded to include planned maintenance and previous exemptions. The only exemption from the new definition is archaeological investigations.

d Additional 'trigger' appointments, for example, the Principal Contractor and Principal Designer are now required for projects where there is more than one contractor working on the site. (Previously triggered by the F10.)

e Increased Client responsibility and additional legal duties.

Commercial clients

CDM 2015 makes a distinction between commercial clients and domestic clients. Client duties apply in full to commercial clients; for domestic clients the duties normally pass to other duty holders.

A 'commercial client' is any individual or organisation that carries out a construction project as part of a business.

Commercial clients have a crucial influence over how projects are run, including the management of health and safety risks. Whatever the project size, the commercial client has contractual control, appoints designers and contractors, and determines the money, time, and other resources for the project.

For all construction projects, commercial clients must:

• Make suitable arrangements for managing their project, enabling those carrying it out to manage health and safety risks in a proportionate way. These arrangements include:

- appointing the contractors and designers to the project (including the principal designer and principal contractor on projects involving more than one contractor) and making sure they have the skills, knowledge, experience, and organisational capability
- allowing sufficient time and resource for each stage of the project
- making sure that any principal designer and principal contractor appointed carry out their duties in managing the project
- making sure suitable welfare facilities are provided for the duration of the construction work
- Maintain and review the management arrangements for the duration of the project
- Provide pre-construction information to every designer and contractor either bidding for the work or already appointed to the project
- Ensure that the principal contractor (or contractor, for single-contractor projects) prepares a construction phase plan before that phase begins
- Ensure that the principal designer prepares a health and safety file for the project and that it is revised as necessary and made available to anyone who needs it for subsequent work at the site

For notifiable projects (where planned construction work will last longer than 30 working days and involves more than 20 workers at any point in the project; or where the work exceeds 500 Person days individual worker days), commercial clients must:

- Notify HSE in writing with details of the project; and
- Ensure a copy of the notification is displayed in the construction site office.

Domestic clients

CDM 2015 makes a distinction between domestic clients and commercial clients, who commission construction work as part of their business.

A domestic client is any individual who has construction work carried out on his or her home, or the home of a family member, that is not done as part of any business. While CDM 2015 places client duties on commercial clients in full, such duties for domestic clients normally pass to:

- The contractor, if it is a single-contractor project, who must take on the legal duties of the client in addition to their own as contractor. In practice, this should involve little more than what they should normally do in managing health and safety risks.
- The principal contractor, for projects with more than one contractor, who must take on the legal duties of the client in addition to their own as principal contractor. If the domestic client has not appointed a principal contractor, the client duties must be carried out by the contractor in control of the construction work.

If a domestic client has appointed an architect (or other designer) on a project involving more than one contractor, he or she can ask that architect or designer to manage the project and take on the client duties instead of the principal contractor. The designer then takes on the responsibilities of principal designer and must have a written agreement with the domestic client, confirming he or she has agreed (as principal designer) to take on the client duties as well as his or their own responsibilities.

Any designer in charge of coordinating and managing a project is assumed to be the principal designer. However, if he they do not have a written agreement with the domestic client to confirm they are taking on the client duties, those duties automatically pass to the principal contractor.

Principal designers

A 'principal designer' is a designer who is an organisation or individual (on smaller projects) appointed by the client to take control of the pre-construction phase of any project involving more than one contractor.

Principal designers have an important role in influencing how risks to health and safety are managed throughout a project. Design decisions made during the pre-construction phase have a significant influence in ensuring the project is delivered in a way that secures the health and safety of everyone affected by the work.

The principal designers must:

- Plan, manage, monitor, and coordinate health and safety in the pre-construction phase. In doing so they must take account of relevant information (such as an existing health and safety file) that might affect design work carried out both before and after the construction phase has started.
- Help and advise the client in bringing together pre-construction information and provide the information designers and contractors need to carry out their duties.
- Work with any other designers on the project to eliminate foreseeable health and safety risks to anyone affected by the work and, where that is not possible, take steps to reduce or control those risks.
- Ensure that everyone involved in the pre-construction phase communicates and cooperates, coordinating their work wherever required.
- Liaise with the principal contractor, keeping them informed of any risks that need to be controlled during the construction phase.

On a domestic-client project, where the domestic client does not appoint a principal designer, the role of the principal designer must be carried out by the designer in control of the pre-construction phase. When working for a domestic client, the client duties will normally be taken on by another duty holder (often the principal contractor on projects involving more than one contractor). However, the principal designer can enter into a written agreement with the domestic client to take on the client duties in addition to their own.

Principal contractors

A 'principal contractor' is appointed by the client to control the construction phase of any project involving more than one contractor.

Principal contractors have an important role in managing health and safety risks during the construction phase, so they must have the skills, knowledge, experience, and, where relevant, organisational capability to carry out this work.

The principal contractor must:

- Plan, manage, monitor, and coordinate the entire construction phase.
- Take account of the health and safety risks to everyone affected by the work (including members of the public), in planning and managing the measures needed to control them.
- Liaise with the client and principal designer for the duration of the project to ensure that all risks are effectively managed.
- Prepare a written construction phase plan before the construction phase begins, implement, and then regularly review and revise it to make sure it remains fit for purpose.
- Have ongoing arrangements in place for managing health and safety throughout the construction phase.

- Consult and engage with workers about their health, safety, and welfare.
- Ensure suitable welfare facilities are provided from the start and maintained throughout the construction phase.
- Check that anyone he appoints has the skills, knowledge, experience, and, where relevant, the organisational capability to carry out the work safely and without risk to health.
- Ensure all workers have site-specific inductions, and any further information and training they need.
- Take steps to prevent unauthorised access to the site; and
- Liaise with the principal designer to share any information relevant to the planning, management, monitoring, and coordination of the pre-construction phase.

When working for a domestic client, the principal contractor will normally take on the client duties as well as their own as principal contractor. If a domestic client does not appoint a principal contractor, the role of the principal contractor must be carried out by the contractor in control of the construction phase. Alternatively, the domestic client can ask the principal designer to take on the client duties (although this must be confirmed in a written agreement) and the principal contractor must work with them as 'client' under CDM 2015.

Further information
- HSE L153 Publication – Guidance on Regulations
- Six CONIAC (HSE – Construction Industry Advisory Committee) Guidance documents.

3
Legislation

3.1 Town and country planning in the UK

Measuring environmental performance
Mike Ridley

Public policy and opinion is directed towards raising environmental standards, and this requires making the performance of buildings and the building process visible to regulators and the marketplace. Measurement systems enable standards to be mandated and give an objective basis for taxation and incentives. They are increasingly a factor in assessing value.

There are three main approaches to measurement and certification:

- Environmental assessment methods
- Energy certification
- Carbon management

Environmental assessment methods

There are a number of different environmental assessment methods used in the UK and internationally. These tools demonstrate the environmental performance of buildings in a measurable way so they can be compared with each other and against prescribed standards. As market awareness has grown they have become benchmarks for quality and value. On the whole they are voluntary, but they are increasingly being mandated by planning policy, funding requirements, the influence of staff, and other stakeholders and market expectations.

The three most common methods used in the UK are BREEAM, LEED, and the SKA rating.

BREEAM

The Building Research Establishment Environmental Assessment Method was introduced by BRE in 1990 and is the market leader in the UK by an overwhelming margin.

Buildings are assessed against criteria in various 'dimensions' of performance and awarded 'credits'. These credits are weighted and aggregated to generate a score. The building's performance is then rated as unclassified, pass, good, very good, excellent, or outstanding. The assessment criteria are periodically reviewed to align with current best-practice standards; thresholds are raised and made more challenging, thereby nudging standards upwards.

There are BREEAM UK schemes tailored for a range of applications.

- **New construction:** a method for assessing all new non-domestic buildings. This has consolidated a number of different schemes specific to different building types.
- **Communities**: a method aimed at integrating sustainable design into the master-planning process for large-scale development.
- **EcoHomes**: a scheme for domestic properties, introduced in 2000 and superseded by the Code for Sustainable Homes (CSH) in 2007.
- **Code for Sustainable Homes**: the government-owned national standard for domestic properties.

- **Refurbishment**: a suite of tools for assessment of domestic and non-domestic refurbishment and fit-out.
- **BREEAM in-use**: introduced to the market in 2009. It allows an organisation to assess the physical characteristics, management and actual environmental performance of a built asset. The aim is to encourage holistic improvements in building use.

For new construction, a BREEAM assessment is undertaken at the completion of the design stage (DS assessment) and a certificate is then issued. The post-construction stage (PCS assessment) of the assessment follows on completion to assure that what was designed is actually built. On conclusion of the project a post-construction certificate is awarded.

The BREEAM methodology awards credits under the following headings:

- **Management**: policy, site management, and process
- **Health and well-being**: all aspects affecting occupants' welfare
- **Energy efficiency**
- **Transport**: location-related factors, including amenities, public transport, and provision of cycling facilities
- **Water consumption**: efficiency and leak detection
- **Materials**: environmental implications and life-cycle impact
- **Waste**: associated both with the construction process and recycling facilities of the completed building
- **Land use**: regarding greenfield and brownfield sites
- **Ecology**: both conservation and enhancement
- **Pollution**: air, water, and the local environment
- **Innovation**: credits which reward new ideas and uptake of new technology

To achieve the highest rating, the optimum approach is to address the main issues at the earliest point of the design process with input from the full project design team. BRE have introduced the accredited professional (AP) qualification for construction professionals able to champion sustainability within a project team. Two additional innovation credits can be awarded if an AP is involved in the project from RIBA Stage 2 Concept Design.

Independent assessors are trained and licensed by BRE. An assessor can be used to coordinate and collate input from the team and to track the development of ideas. The assessor can also give advice about BREEAM to the entire project team at the start of the project. As the scheme progresses, the assessor can:

- provide specialist advice on the specification of products to achieve particular BREEAM credits;
- undertake preliminary BREEAM assessments to assess the predicted rating; and
- provide a sustainability report for submission for planning approval.

For further information visit www.breeam.com

LEED

Leadership in Energy and Environmental Design (LEED) is an American environmental system developed by the US Green Building Council. Reflecting US economic influence, it is the dominant system worldwide and has some penetration in the UK, particularly for US and multinational corporations adopting common standards across their portfolio.

There are five rating systems that address multiple project types:

- **Building design and construction**
- **Homes**
- **Neighbourhood development:** applies to larger development projects

- **Interior design and construction**
- **Building operations and maintenance:** applies to existing buildings in use

LEED credits for new buildings are awarded for performance in different 'dimensions' of sustainability, similar to BREEAM but with a slightly different emphasis. Ratings awarded are Certified, Silver, Gold, and Platinum.

For details of operation of the system and gaining certification visit www.usgbc.org/.

SKA rating

SKA rating began life as a research project commissioned in 2005 by Skansen Ltd, the RICS and AECOM to establish metrics for the environmental impact of an office fit-out. It was launched as a product in 2009. Since then, retail and higher education versions have been introduced.

The online SKA tool is freely available to carry out an informal self-assessment of a project. A formal assessment requires the involvement of a licensed SKA assessor.

SKA comprises 104 good-practice measures covering similar categories to the other assessment systems. In reflection of the varying scope of fit-out projects, the SKA rating scores only those measures that are relevant to the project ('Measures in Scope'). Typically, 30–60 measures are likely to apply to most projects, and they are ranked depending on the importance from a sustainability perspective. The project has to achieve a number of the highest-ranked Measures in Scope in order to rate. These are known as Gateway Measures.

The SKA rating can be assessed three times during the life cycle:

- during design

- at handover
- during occupancy

The score is ranked according to the achievement of three thresholds, Gold, Silver, and Bronze (75%, 50% or 25% respectively).

For further information on the operation of the system and gaining certification visit www.rics.org/.

Systems used overseas

LEED is the dominant system worldwide and BREEAM is widely used in Europe. Other systems that have been developed in specific jurisdictions include:

DGNB Certification System
The German Sustainable Building Certification was developed by the German Sustainable Building Council (DGNB) together with the Federal Ministry of Transport, Building, and Urban Affairs (BMVBS) to be used as a tool for the planning and evaluation of buildings throughout the whole life cycle.

Green Star
This is an environmental labelling system developed by the Green Building Council of Australia, based on similar principles to both BREEAM and LEED. It is a voluntary system generally used in the Australasian region and South Africa.

IGBC Green Rating System
Indian Green Building Council (IGBC) has developed voluntary green building rating programmes (Homes, Special Economic Zones [SEZ], and Factory). Rating programmes help projects to address all aspects related to environmental issues and to measure the performance of the building/project. The rating system evaluates certain credit points using a prescriptive approach and other credits on a performance-based approach.

CASBEE

Comprehensive Assessment System for Building Environmental Efficiency (CASBEE) is a voluntary system in Japan, but is currently being included in planning policy in some regions. In contrast to the other systems, CASBEE looks at Q – Quality and performance (which evaluates improvement in living amenity for the building users, within the boundary of the building) and L – Loadings (which evaluates negative aspects of environmental impacts of the building).

Energy certification

Environmental assessment methods measure building performance using a weighted 'basket' of disparate factors. Other management systems focus primarily on resource use and particularly energy consumption. This is a simpler approach and is favoured by regulators for that reason.

Energy Performance Certificate (EPC)

This requirement was introduced in 2008 as a direct consequence of the Energy Performance of Buildings Directive which is EU legislation. An EPC is a measure of the designed energy usage, and must be produced when a building is constructed, sold, rented out, or subject to major alterations. It provides the property with an asset rating on a scale from A through G, and must be accompanied by recommendations on improving the energy performance of a building. There is no requirement that these be implemented.

Display Energy Certificate (DEC)

DECs are required to be displayed in large buildings and those occupied by public authorities and certain public institutions. They are based on actual energy usage and are intended to reveal how well buildings are performing. DECs must be reviewed annually.

Passivhaus

'Passivhaus' is an ultra-low energy standard for houses originating in Germany, but now applies quite widely to projects throughout Europe. It relies on passive design (solar orientation, thermal mass, and so on), very high levels of insulation and high standards of workmanship. The standard has simple but demanding threshold requirements:

- Demand for space heating and cooling of less than $15kWh/m^2$/year
- Total primary energy consumption for all appliances, domestic hot water, space heating and cooling of less than $120kWh/m^2$/year
- Fewer than 0.6 airchanges/hr

The Passivhaus standard is seen as representing a more durable and occupier-proof solution to reducing energy consumption than design approaches that rely on technology which can be prone to failure or incorrect operation.

Carbon management

The main driving force for the international sustainability agenda is the imperative to control and reduce man-made carbon emissions. This is closely linked to, but not identical to, energy consumption. In 2008 the UK government passed into law the Climate Change Act making a legal commitment to reduce carbon emissions (at 1990 levels) by 80% by 2050. The Act requires the establishment of a carbon budgeting system to demonstrate progress towards this target, setting emissions caps for 5-year intervals.

As buildings (through their construction, occupation, and the processes that take place in them) contribute 48% of the nation's CO_2 emissions, the built environment, and how it is used, has a fundamental role to play in meeting the targets.

The government has committed that all new homes should be "zero carbon" by 2016 with new non-domestic buildings following suit by 2019.

"Zero carbon" is defined as the mitigation, through various measures, of all the carbon emissions produced by 'regulated' energy use at a particular site. Regulated use includes space heating and cooling, hot water, and fixed lighting (as outlined in Part L1A of the Building Regulations) but not cooking and plug-in appliances.

Carbon is moving into a role of critical resource traditionally occupied solely by finance. To ration its use, similar techniques have evolved for carbon 'footprinting', planning and whole-life modelling to be validated by independent auditing and certification. To encourage uptake, a carbon plan is a necessary precondition for investment grants from Central Government.

Embodied carbon

The 'carbon footprint' from the construction and use of a building includes all stages in its lifecycle:

- Design
- Materials or product manufacture
- Distribution
- Assembly on site
- In use
- Demolish or refurbish

As can be seen from the following table, the overwhelming bulk of carbon emissions arise 'in use' through the occupation of buildings by people. There is, however, also a sizable component (16%) arising from the manufacture and distribution of materials and their incorporation into buildings on site.

Table 3.1 Estimate of the carbon footprint of UK construction 2008 (Source BIS 2010).

Subsector	Greenhouse Gas Emissions Mt CO2e	% of total carbon footprint
Design	1.3	0.5%
Manufacture	45.2	15%
Distribution	2.8	1%
Operations on-site	2.6	1%
In Use	246.4	83%
Refurb/demolition	1.3	0.4%
Carbon Footprint Total	**298.4**	**100%**

Current regulation only applies to designed 'In Use' performance of the building that is the carbon emitted by people in occupation. The trend for future regulation is likely to include the carbon emitted through the construction of the building and subsequent refurbishment and demolition – the 'embodied carbon'. It has been estimated that the embodied carbon for a typical home is equivalent to the sum of emissions over the first 10 years in use.

The incorporation of embodied energy shines the spotlight on sourcing and manufacturing processes. Measurement methodologies for calculating and certifying the embodied carbon of a given material or prefabricated component are being developed.

Environmental Management Systems

EMSs are means by which a company can ensure and demonstrate that environmental management is a core value and embedded in its working practices.

ISO 14001
ISO 14001 is an international standard that sets out how an organisation can introduce an effective Environmental

Management System (EMS) and manage its environmental impacts. It is not specific to property, construction and development but is generically applicable to all forms of economic activity. The standard is designed to address the balance between maintaining profitability (i.e. economic sustainability) and reducing environmental impact.

3.2   Landlord and tenant

Dilapidations – England and Wales
Alistair Cooper

The term 'dilapidations' refers to breaches of lease obligations, either express or implied, and usually relates to reinstatement, repair, decoration, breaches of statute or other specific requirements, and associated costs.

The fundamental purpose of a 'Schedule of Dilapidations' is to identify the breaches of covenant. The legal remedy for the breaches is typically a claim for damages. Damages are compensatory and not intended to penalise the guilty party. They are intended to compensate an innocent party for any loss which is caused by the breaches of covenant.

A Schedule of Dilapidations is usually prepared as part of a wider claim which itself should represent the damage actually suffered by the landlord, and which is capped under Section 18(1) of the Landlord and Tenant Act 1927, as well as by common law principles. (See Section 18(1) of the Landlord and Tenant Act [1927].)

Types of Schedule

Schedules of Dilapidations commonly fall into the following categories:

- **Interim**: served during the term, and not in anticipation of the end of the term
- **Terminal**: served shortly prior to, or after, the end of the term

Guidance

The RICS guidance note has been drafted having regard to the *Civil Procedure Rules,* Binding Practice Direction, and the adopted Dilapidations Protocol. Currently in its sixth edition, the guidance note is not compulsory but does represent what the RICS considers to be best practice. The surveyor is however cautioned that the guidance note is not appropriate in all cases.

Property Litigation Association (PLA)

The PLA 'Pre-action Protocol for Damages in Relation to the Physical State of Commercial Property at the Termination of a Tenancy' (The Dilapidations Protocol) was first introduced in 2002. The aim of the protocol is to avoid litigation, encourage the early and full exchange of information between parties, and to discourage landlord's spurious or inflated claims.

The Dilapidations Protocol was formally adopted by the Civil Justice Council (CJC) on 14 October 2011 and commenced in April 2012. The Protocol should be read in conjunction with the Practice Direction under the Civil Procedure Rules (CPR). The Dilapidations Protocol encourages parties to be fair and open in their negotiations with the emphasis on avoiding litigation. Any party who does not follow the Protocol will be severely criticised if the claim reaches Court.

The RICS and PLA guidance are in broad agreement and key issues to note are:

- recommended reasonable response times (generally within 56 days);
- standardisation of various procedures;
- emphasis on the requirement for the initial claim to be reasonable;
- requirement for a formal Diminution Valuation only prior to issue of proceedings;
- consideration of Alternative Dispute Resolution;
- consideration of the role of the surveyor as either 'adviser' or 'expert';
- consideration of the assessment of loss; and
- endorsements required by both landlord and tenant surveyors.

Stages in the dilapidations process

There are various stages in the dilapidations process. These can be broken down into five main parts, as follows:

Stage 1 – preparation

Obtain and appraise all relevant documentation including (but not limited to):

- leases;
- licences to alter;
- schedules of condition;
- side letters;
- heads of terms
- photographs;
- fit-out specifications;
- agents' letting brochures;
- any statutory notices served;
- deeds of variation;

- schedules of landlord's and tenant's fixtures and fittings;
- details of outstanding service charges; and
- any rent deposit agreements.

Stage 2 – inspection

This stage must be comprehensive and thorough. Increasingly complex buildings-specialist input, such as that provided by a structural engineer or mechanical or electrical consultant, may be required.

The surveyor should establish the original condition at the beginning of the term (if possible) and the standard of repair that the tenant has covenanted to achieve, taking into account the age, character, and locality of the premises when let (see *Proudfoot v Hart* [1890]).

In particular the surveyor should:

- identify items of disrepair;
- check if the item falls within the demise;
- check if the item is subject to the covenant or covenants;
- consider if the item is out of repair (i.e. has there been a change from a previously better condition);
- consider the nature of remedial work that is reasonably necessary;
- take measurements to aid calculation of the cost of the remedial works and use as proof (to substantiate costs) as required at a later date;
- consider whether statutory legislation is to be complied with; and
- consider the impact of 'green' lease clauses.

Stage 3 – preparation of the Schedule of Dilapidations and claim

The example below shows a typical format for a Schedule of Dilapidations. The schedule should be set out in a logical order and broken down by on an itemised basis. The initial schedule document should contain information in the first eight columns. Once negotiations commence, additional columns are added to provide tenant's comments and their own cost assessment. Further columns may be necessary for both sides as subsequent negotiations progress. The combined document is commonly referred to as a 'Scott Schedule'.

The Schedule of Dilapidations should be accompanied by a claim letter, which must include:

- the name and address of the landlord and the tenant;
- a clear summary of the facts on which the claim is based;
- the Schedule of Dilapidations (a separate document);
- any documents relied upon, such as invoices and evidence of costs and losses;
- confirmation that the landlord and advisers will attend meetings;
- a date by which the tenant should respond;
- a summary of the claim including: cost of works, preliminaries, overheads and loss of profit, surveyors' fees for preparing the Schedule (quantified and substantiated), loss of rent, loss of service charge, surveyors' fees for negotiating a settlement (projected), and any sums paid to a superior landlord; and
- a diminution valuation (depending on the landlord's intent).

The Schedule of Dilapidations may be served within a reasonable time prior to the termination of the tenancy, and generally not more than 56 days afterwards, though the landlord may have up to 6 years before it may commence proceedings. If the schedule is served prior to the end of the term, then the landlord should, at termination, confirm the schedule or issue a further schedule.

A notice for the reinstatement of alterations may have to be served upon the tenant prior to the end of the tenancy. It is essential that the terms of any reinstatement of alterations are understood prior to the end of the lease. Some of these types of covenants may require prior notice to be valid.

The terms of any reinstatement of alterations covenants need to be established prior to the end of the tenancy. These covenants can, sometimes, require notice to be served by landlords within a reasonable time period prior the end of the term. If the landlord fails to serve such requisite notice, then any subsequent request for the reinstatement of alterations, following the end of term, may be invalid.

Electronic copies of the Schedule of Dilapidations should be provided to facilitate easier negotiation, preferably in a Scott Schedule format as shown above.

Stage 4 – the Quantified Demand

The Dilapidations Protocol introduced the concept of the 'Quantified Demand' which should be issued by the landlord to the tenant within a reasonable time (usually 56 days after termination of the tenancy). The Quantified Demand is similar to a 'summary of claim' but includes additional information which the surveyor may not be party to. Items within the Quantified Demand include the following:

- details of the landlord, tenant, and the property;
- the lease details;
- the position regarding recovery of VAT;
- a summary of known facts; including the landlord's intentions for the property;
- a summary of monetary sums, including, but not exclusively, the Schedule of Dilapidations;
- details of the supporting documents on which the claim is founded;

Table 3.2

Item No.	Clause No.	Breach complained of	Remedial work required	Landlord's cost for			Landlord's sub-total cost	Tenant's response [date]	Tenant's cost [date]
				Reinstatement	Repair	Redecoration			

- the Quantified Demand, including a time period for response; and
- confirmation that the landlord(s), or their advisors, are happy to attend meetings.

Stage 5 – the response and negotiations

Following submission of the claim, the tenant should respond within a reasonable time period, usually 56 days after the landlord sends the Quantified Demand. However, in respect of interim dilapidations, the time period for response required by statute, or under the lease, may be considerably shorter if the tenant's position is to be protected.

The pre-action protocol recommends that the landlord and tenant or their respective surveyors meet prior to the tenant being required to respond to the Quantified Demand and generally within 28 days after the tenant sends their response. The meetings are to be held on a 'without prejudice' basis and the parties are intended to try and agree as many items in dispute as possible.

The pre-action protocol also advises that prior to litigation both parties should consider some form of alternative dispute resolution procedure and the landlord should quantify their loss. The landlord(s) can quantify their loss by either providing a formal diminution valuation or an account of any actual expenditure in respect of the breaches and consequential losses.

Interim dilapidations claims

These claims are made during the lease term and are typically a consequence of one of the parties becoming concerned with breaches of the lease obligations by the other. This is a complicated area of law, so legal advice should always be taken on the appropriate drafting, service, and timing of notices and counter-notices.

Interim dilapidations claims by a tenant may also seek the following remedies: specific performance, damages, the right to undertake works, offset against rent, or repudiation and quitting of the premises. The tenant's options will, however, differ in each case and should be considered carefully.

It is usually the landlord who will exercise the right to serve an interim schedule of dilapidations. These schedules are typically subject to one of three remedies. These are as follows: forfeiture, damages, or specific performance. Each option has its own advantages and disadvantages with statutory requirements imposed in certain circumstances requiring notices to be served under Section 146 of the Law of Property Act 1925. Such notices can lead to statutory defences from the tenant and requisite counter-notices.

A landlord would usually need to satisfy one or more of the following five grounds (as detailed within the Leasehold Property Repairs Act 1938):

- that immediate remedying is necessary to protect the value of the landlord's interest in the property, or that the value has already been affected;
- that the work is necessary to comply with legislation;
- that where the tenant does not occupy the whole building, the work is necessary in the interests of other occupiers;
- that immediate remedying of the defect will be substantially cheaper than would be the case if the work was delayed; or
- special circumstances exist.

Specific legal advice, together with advice from surveyors, should be sought by landlords or tenants looking to pursue, or defend, an interim schedule of dilapidations having regard to the particular facts and circumstances.

Jervis v Harris notices (repairs notices)

A repairs notice, commonly known as a 'Jervis v Harris notice' (see *Jervis v Harris* [1996]), refers to a specific covenant sometimes contained within a lease. This covenant enables a landlord to give notice to a tenant highlighting breaches of repair during the term. The intention is to require the tenant to remedy these breaches. If the tenant fails to carry out the works, the landlord is entitled to enter the premises and undertake the works as notified. The landlord may then recover the cost of doing so, from the tenant, usually as a debt rather than damages avoiding the statutory cap of Section 18(1) of the Landlord and Tenant Act 1927 and common law.

The lease covenant will set out a series of steps that the landlord must comply with prior to having the option to enter the premises, carry out the works, and recover the cost of doing so. Typically the covenant will specify the exact requirements of the service of the repairs notice and importantly specific steps which are to be taken including prior to any initial inspection of the demise by the landlord. The covenant will also detail the time period in which the tenant must comply with the notice.

The notice itself is only to include breaches of covenant allowed for by the clause and no others. It should also not identify remedial works, as there are often a number of ways of remedying a breach. The tenant (being the party responsible, under the lease, to remedy the breaches) may carry out the work in a different manner than the landlord may envisage but to a standard compliant with the lease covenant.

Legal advice should always be obtained throughout this process, and caution is required from the outset. Significant problems and restrictions could be encountered (see *Creska Ltd v Hammersmith & Fulham LBC* [1998]) including trespass by the contractor (leading to a counter claim for damages), access to power and water, interference with the tenant's quiet enjoyment, and an adverse effect on the relationships between the parties.

Break clauses

Leases increasingly include an option for the landlord or the tenant to bring the lease to an end before the contractual expiry date. It is important to establish whether the break clause is 'condition precedent' or not.

Where the clause is conditional, the party exercising the break may be required to comply with a variety of conditions before the break can be deemed effective. Compliance may include absolute compliance with all obligations within the lease and/or requiring that vacant possession be provided at the break date.

The risk of non-compliance with a conditional break clause is often substantially greater than the cost of the works necessary to comply. It is therefore essential to fully understand and adhere to the lease requirements. The following cases shed some light as to how break clauses might be interpreted by the courts.

- *Fitzroy House Epworth Street (No 1) Ltd v The Financial Times Ltd* [2006]
    - Test of 'material compliance' of a conditional break clause
    - Requirements of landlords and their advisers to assist tenants under conditional break clause circumstances
- *Legal and General Assurance Society Ltd v Expeditors International (UK) Ltd* [2006]
    - Financial settlements in respect of dilapidations associated with break clause preconditions and vacant possession
- *NYK Logistics (UK) Ltd v Ibrend Estates BV* [2011]
    - Defined the meaning of 'vacant possession'

A raft of other case law is ever developing on the subject of break clauses and it is recommended that in every case legal advice should be sought at the earliest possible opportunity to ensure the correct operation of the break option.

VAT and dilapidations

VAT in respect of dilapidations is a complex subject with its own set of rules, regulations, and case law. VAT does not directly apply to dilapidations. A dilapidations settlement payment is not a 'taxable supply' for the purposes of VAT, as the payment is deemed to be one of damages and not a supply payment (see HMRC VAT Notice 742: land and property [2012]). However, there are situations where a sum can be paid in place of VAT. The Dilapidations Protocol, Clause 4.2.1, requires that the recoverability of VAT should be considered and stated within the quantified demand.

The Finance Act 1989 introduced major changes that included giving UK payers of VAT the option to pay VAT on supplies relating to an interest in commercial land.

Where a person or company is registered for VAT there is a special statutory exemption to the charging of VAT on such supplies. Taxpayers can waive this exemption if they desire by notifying HMRC. If a landlord has elected to waive his or her exemption on a building, he or she must charge VAT on such items as rent received, and it will not be appropriate to include VAT as part of a dilapidations claim (the landlord being able to offset this VAT received on the rent against business-related expenses, such as VAT on materials and labour for works undertaken). If the VAT exemption has not been waived, then the landlord will be unable to offset the VAT and it may form part of a dilapidations claim.

Once the landlord's VAT position has been established, a simplified VAT analysis might follow the steps shown in the diagram below.

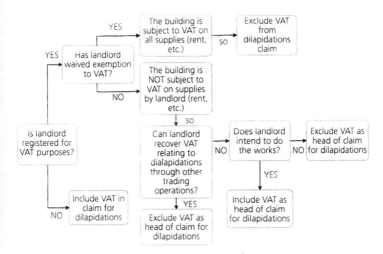

Figure 3.1

Summary of important statutes

- Landlord and Tenant Act 1927, Section 18(1)
  - Limits the cost of a claim for breach of covenant.
- Law of Property Act 1925, Section 146
  - Prescribes the form of notice for re-entry for forfeiture.
- Law of Property Act 1925, Section 147
  - Provides relief for tenants on long leases in respect of internal redecorations.
- Leasehold Property (Repairs) Act 1938
  - Gives protection to certain tenants in respect of Section 146 notice above.
- Landlord and Tenant Act 1954, Part IV
  - Extension of Leasehold Property (Repairs) Act 1938.
- Defective Premises Act 1972
  - Provides that a landlord shall be liable for lack of repair in cases where the landlord knew or ought to have known of the defect.
- The Civil Procedure Rules 1998

- Introduced by Lord Woolf, they provide rules and practice directions for dispute procedures.
- The Fraud Act 2006
  - The injured party need not suffer loss, implications for practitioners in preparing and defending claims.

Summary of important case law

There have been a number of leading decisions relating to dilapidations law in the last few years. The following cases give an indication of developing areas of law.

- *Scottish and Mutual Assurance Society Ltd v British Telecommunications plc* [1994] (unreported)
  - Section 18(1) of the Landlord and Tenant Act 1927 Part II
  - Loss of rent
  - Notice for reinstatement of alterations
- *Shortlands Investments Ltd v Cargill plc* [1995] EGLR 51
  - Section 18(1) of the Landlord and Tenant Act 1927 Part II
- *Trane (UK) Ltd v Provident Mutual Life Assurance Co Ltd* [1995] WGLR 78
  - Compliance with conditions of break clauses
- *Jervis v Harris* [1996] EGLR 78
  - Use of provision for landlord's re-entry
  - Extent of recovery of expenditure
- *Creska Ltd v Hammersmith and Fulham LBC* [1998] EGLR 35
  - Landlord's right of re-entry
- *Mannai Investments Co Ltd v Eagle Star Life Assurance Co Ltd* [1997] EGLR 69
  - Accuracy of notice showing intent to break tenancy
- *Credit Suisse v Beegas Nominees Ltd* [1994] 4 All ER 803
  - Establishing the interpretation of the repairing covenants and the different types of obligation placed on the tenant

- *Fitzroy House Epworth Street (No 1) Ltd and Fitzroy House Epworth Street (No 2) Ltd v The Financial Times Ltd* [2006] CA1 2207
    - Test of 'material compliance' of a conditional break clause
    - Requirements of landlords and their advisers to assist tenants under conditional break clause circumstances
- *Legal and General Assurance Society Ltd v Expeditors International (UK) Ltd* [2006] EWHC 1008
    - Financial settlements in respect of dilapidations associated with break clause preconditions and vacant possession
- *Joyner v Weeks* [1891] 2 QB 31
    - Common law measure of damages
- *Latimer v Carney* [2006] EWCA Civ 1417
    - Measure of damages, Section 18(1) of the Landlord and Tenant Act 1927.
- *Janet Reger International v Tiree Holdings* [2006] EWHC 1743 (Ch D)
    - Liability for disrepair, deterioration from a previous better condition
- *Riverside Property Investments Ltd v Blackhawk Automotive* [2005] 01 EG 94
    - Standard of repair and recovery of costs
- *Carmel Southend Ltd v Strachan & Henshaw Ltd* [2007] EHWC 1289 (TCC); (2007) 35 EG 136
    - Standard of repair, patch repair sufficient, tenant not required to put premises into perfect repair or pristine condition.
- *Lyndendown Ltd v Vitamol Ltd* [2007] EWCA Civ 826; (2007) 47 EG 170
    - Damages not recoverable if subtenant remains in occupation under the same terms
- *Prudential Assurance Co Ltd v Exel Ltd & Another* [2009]
    - The importance of serving the notice on the correct party
- *PGF II SA v Royal & Sun Alliance* [2010] EWHC 1459 (TCC)

- The award of costs as damages and clarification of supersession and the landlord's intentions
- *PCE Investors Ltd v Cancer Research UK* [April 2012]
  - The importance of fully complying with the lease requirements, without deduction or offset.
- NYK Logistics (UK) Ltd v Ibrend Estates BV [2011] EWCA Civ 683
  - The meaning of 'vacant possession'

Green leases

Businesses have increasing requirements to reduce their carbon footprint, resulting in a new breed of 'green lease' coming into being which addresses both energy usage and sustainable construction methods. Green leases are largely untested with regard to dilapidations but are likely to impact on the way both terminal and interim dilapidations schedules are compiled in the future. Examples of the additional dilapidations considerations of green leases include:

- works required to restore, or indeed improve, an EPC rating to its former level – this may be particularly contentious in coming years as EPC ratings gain additional legal standing with regard to the ability to sell or let a property, or as computer models become more sophisticated;
- does the condition of the property affect energy usage – this may be particularly relevant for interim schedules;
- the cost of repairs using environmentally sourced materials may be considerably higher than standard products;
- the extent of reinstatement required at the end of the term – for example, should a tenant be required to remove wind turbines that represent a high maintenance liability, but which may significantly improve the property's environmental rating;
- the extent of renewal (as opposed to repair) required at the end of the term; and

• the cost, or benefit, of recycling materials removed from the building.

No doubt further clarification and guidance, both from professional and legal sources, will be required.

Dilapidations – Scotland
Gordon Hamilton

At a practical surveying level, dilapidations work in Scotland differs very little from that in other parts of the UK. There are, however, a few significant differences in the legal principles and the statutory background to dilapidations work in Scotland, all of which are potential pitfalls for the unwary.

In Scotland, unlike England and Wales, there is little statute or case law regulating commercial leases and their effect. As a result, the liability for repair of a commercial property, under a contract of lease, is principally determined by Scots common law, and the extent to which this has been displaced by the wording of the conditions set out in the lease.

It is therefore essential to understand the legal relationship between the landlord and tenant under Scots common law, any relevant statute law, and, in particular, that imposed by the contract of lease.

At Scots common law, a landlord provides the tenant with an implied warranty that the leased premises are wind and watertight and are reasonably fit for their purpose at the outset of the lease. The landlord also has a continuing duty to keep the premises in a tenantable condition and wind and watertight during the currency of the lease. The tenant, therefore, has very little liability for repairs at common law. The tenant is expected to use reasonable care in the management of the premises and is only liable for any damage attributable to his or her negligence.

Most modern leases of commercial property will transfer some or all of the landlord's common law repairing liability onto the tenant. Where doubt or ambiguity exists over the meaning of the wording of the lease, as to where the repairing obligations may lie, then the courts will apply these common law principles. This may have the effect of leaving the obligation with the landlord unless it can be otherwise clearly shown to be that of the tenant.

Scots common law is also derived from the limited number of decisions in cases heard in the Scottish courts. Decisions from courts in other parts of the UK, although persuasive, are certainly not treated as binding precedents in Scotland.

There are few statutory provisions that affect the Scots common law position outlined above or, more importantly, the obligations set out in the contract of lease. This is significantly different to the situation in England and Wales.

Scotland has no equivalents to the Landlord and Tenant Acts of 1927, 1954, and 1988, the Leasehold Property (Repairs) Act 1938, the Law of Property Act 1925, or the Defective Premises Act 1972. Equally importantly, the Civil Procedure Rules do not apply in Scotland. The third edition of the Property Litigation Association's so-called 'Dilapidations Protocol' published in May 2008, has no legal significance in Scotland. The RICS *Dilapidations* guidance note states in its introduction that it applies only to England and Wales. However, Scotland now has its own RICS guidance note: *Dilapidations in Scotland*, the RICS guidance note (2nd edition), which was published in January 2011.

There are, however, many other statutes and statutory instruments which may have an influence upon commercial property leases in Scotland, including:

- the Occupiers Liability (Scotland) Act 1960;
- the Factories Act 1961;
- the Offices, Shops and Railway Premises Act 1968;

- the Environmental Protection Act 1990;
- the Clean Air Act 1993:
- the Equality Act 2010;
- the Fire (Scotland) Act 2005; and
- the Control of Asbestos Regulations 2012.

These and many others have, however, only an indirect bearing on the relationship of the landlord and the tenant in Scotland, and there are therefore very few statutory implications that have to be taken into account when interpreting Scottish commercial leases. Accordingly, in almost all cases relating to commercial property in Scotland, the text of the lease can largely be taken to mean what it says, subject to the overriding requirements of the common law.

The terms 'interim' and 'terminal' schedules of dilapidation have now become widely used in Scotland. They are borrowed from England and Wales. These terms have no legal relevance under Scots law and as a general rule their use is not essential.

As with the practice in other parts of the UK, there are differences between what might be found in a schedule served during the currency of a lease (an interim schedule) and one served at or near termination of the lease (a terminal schedule).

A schedule served during the currency of the lease and some time before expiry is normally limited to requiring the tenant to remedy what is alleged as a breach of a repairing obligation or some other specific obligation of the lease. Unlike in England and Wales, there is no limit imposed on the nature, extent or significance of the defects raised in an interim schedule, or their cost in relation to the value of the property. The only test is whether the tenant is in breach of its repairing obligation. In preparing an interim schedule, however, the landlord would not do its cause much good in being overzealous by including insignificant items that are of limited relevance to the landlord's interest in the property at that time.

A schedule served at or near expiry of the lease tends to be a much more comprehensive document, and would include not only current breaches of a repairing obligation but also might forewarn the tenant of other breaches for which the tenant will be held liable at expiry. In addition, redecoration, minor repairs, and the reinstatement of the premises by removal of tenants' fixtures and fittings or alterations would all normally be included, where these are relevant.

In default of the tenant, the main remedies open to the landlord under Scots law are as follows:

- The landlord can apply to the courts for a '**decree of specific implement**'. In this case the courts issue an order requiring the tenant to comply, or, alternatively, give the landlord authority to enter the premises and carry out the work at the cost of the tenant. This remedy is mainly used in circumstances where the lease still exists rather than where it has already terminated.
- If the lease makes available such a provision, the landlord can lawfully **enter the premises**, usually after a stipulated period of time, and carry out the works itself and recover any costs from the tenant as a debt.
- The landlord can **terminate or 'irritate' the lease** through the irritancy clause in the lease (assuming there is one), usually on the grounds of breach of contract. The landlord may then sue the tenant for damages. This remedy is, by its nature, only of relevance during the period of the lease and is subject to the provisions of the Law Reform (Miscellaneous Provisions) (Scotland) Act 1985.
- At termination of the lease, if the tenant has failed to honour its repairing obligation, then the landlord might claim a sum in lieu or, if all else fails, **sue the tenant for damages** for breach of contract.
- There may of course be **alternative dispute resolution** (ADR) provisions within the lease which can be used in lieu of pursuing a claim for damages through the courts.

So what is the measure of damages to the landlord in the event of tenant default? There is no definitive answer to this question in Scotland. In England and Wales, statute and the 'diminution in value' approach limit the amount of damages recoverable. There is no such limit on the damages recoverable in Scotland. A more pragmatic approach is needed, and there may indeed be more than one method of quantifying the measure of damages.

It is often said that the 'diminution in value' is the English approach, and the 'cost of repairs' the Scottish approach. This is an oversimplification. The courts in Scotland have confirmed that a landlord is 'entitled' to quantify the claim in the first instance by reference to the cost of repairs, but that it is open to the tenant to prove the 'actual loss' to the landlord is less than that, perhaps measured by some other means. The burden of proof in the first instance to prove loss is that of the landlord.

Tacit relocation

In Scotland, with the absence of a statutory right of lease renewal (there are a few exceptions), the common law doctrine of tacit relocation applies. In simple terms, in a situation where a landlord or tenant fails to give notice to quit within the appropriate timescale, then the lease will continue for a period of 1 year or the period of the lease if it was for less than 1 year. All of the lease terms remain in force, including the repairing obligations.

Case law

There are a number of cases which clarify key elements of dilapidations and repair in Scotland. These can be summarised as follows:

- *Turner's Trustees v Steel* [1900] 2F 363: The implied warranty of fitness for purpose
- *Blackwell v Farmfoods (Aberdeen) Ltd* [1991] GWD 4-219: The landlord's implied warranty at common law of fitness for purpose in relation to common parts
- *Hastie v City of Edinburgh District Council* [1981] SLT 61 & 92: Inadequate description of the premises
- *Marfield Properties v Secretary of State for the Environment* [1996] SCLR 749: The common Law of the Tenement does not apply in leases to assist in determining what would or would not be common parts
- *Taylor Woodrow Property Co. Ltd v Strathclyde Regional Council* [1996] GWD 7-397: A comprehensive examination of the language of a repairing clause
- *Thorn EMI Ltd v Taylor Woodrow Industrial Estates Ltd* [1982], unreported: Wording sufficient to extend liability of tenant to include extraordinary repairs
- *Cantor's Properties (Scotland) Ltd v Swears & Wells Ltd* [1980] SLT 165: The importance which the courts place on the interpretation of words in the context in which they appear in the lease
- *West Castle Properties Ltd v Scottish Ministers* [2004] SCLR 899: Replacement of plant and equipment
- *Scottish Discount Co. Ltd v Blin* [1985] SC 216: Identified the proper tests to determine whether an item is or is not a heritable fixture
- *Cliffplant Ltd v Kinnaird; Mabey Bridge Co. Ltd v Kinnaird* [1982] SLT 2: Tenant's fixtures and fittings
- *House of Fraser Plc v Prudential Assurance Co. Ltd* [1994] SLT 416: Landlord's obligation to repair and service charges
- *Kidneat Ltd v N.C.R. Ltd,* unreported: Common repairs
- *Lowe v Quayle Munro Ltd* [1997] GWD 10-438: Landlords to carry out work in a 'fair and reasonable' manner
- *Mars Pension Trustees Ltd v County Properties & Developments Ltd* [1999] SC 267: Exclusion of landlord's implied warranty at common law on both the leased premises and any common parts to be made in the clearest terms

- *Prudential Assurance Co. Ltd v James Grant & Co. (West) Ltd* [1982] SLT 423: This case established that the absolute rule for the measure of damages in the English case *Joyner v Weeks* was not parts of Scots Law. It also confirmed that the ceiling on damages (i.e. the diminution in value of the landlord's reversionary interest) was inapplicable in Scotland.
- *Admiralty v Aberdeen Steam Trawling & Fishing Co.* [1910] SC 553: Landlord must take (even if in fact he has not) all reasonable steps to minimise his loss.
- *Euro Properties Scotland Ltd v Alam* [2000] GWD 23-896: Irritancy and the principles of the 'fair and reasonable landlord'
- *Co-operative Insurance Society Limited v Fife Council* [2011] and *Crieff Highland Gathering Limited v Perth and Kinross Council* [2011]: Both relating to the extent of the parties' liabilities, in particular relating to 'extraordinary repairs'

Dilapidations – Northern Ireland

Harry Dowey

The foundation of dilapidations law in Northern Ireland is the Conveyancing Act 1881. In addition, the Business Tenancies (Northern Ireland) Order 1996 may be held as applicable in certain circumstances. The Law of Property Act 1925, Landlord and Tenant Act 1927 and the Leasehold Property Repairs Act 1938 do not extend to Northern Ireland.

In practice, the current RICS Dilapidations guidance largely sets the protocol for the execution of dilapidations work in Northern Ireland. One important difference between Northern Ireland and England and Wales is that diminution valuation is not an alternative remedy, because the absence of specific legislation in Northern Ireland means that a working representation is applied that is self evident that the cost of repairs that are due to be done equals the diminution in the value of the reversion.

Significantly, unlike England and Wales, Northern Ireland has, to date, no recorded case law on the subject of dilapidations.

Section 18(1) of the Landlord and Tenant Act 1927
Alistair Cooper

Section 18(1) of the Landlord and Tenant Act 1927 may cap a landlord's common law claim for damages for breach of covenant at the end of the term. Section 18(1) is divided into two limbs.

The first limb provides that:

> *Damages for a breach of a covenant to keep or put premises in repair during the currency of a lease, or to leave or put premises in repair at the termination of a lease, whether such covenant or agreement is expressed or implied, and whether general or specific, shall in no case exceed the amount (if any) by which the value of the reversion (whether immediate or not) in the premises is diminished owing to the breach of such covenant or agreement as aforesaid.*

> *©Crown copyright material is reproduced under the Open Government Licence v1.0 for public sector information: www.nationalarchives.gov.uk/doc/ open-government-licence/*

The first limb of Section 18(1) applies in all dilapidations cases within England and Wales. The common law claim in dilapidations usually comprises the following components:

- cost of works;
- fees for preparation and service of the Schedule;

- fees for carrying out of the works;
- fees for negotiating the claim;
- time related losses (e.g. loss of rent, loss of empty rates, loss of insurance and loss of service charge) and;
- VAT.

The first limb of Section 18(1) stipulates that regardless of the amount of the common law claim, the landlord's entitlement in damages is limited to the amount by which the value of its interest has been diminished owing to the breach of the tenant's covenant to repair. This is not the same as the covenant to redecorate or to reinstate.

Section 18(1) relates to repair only, and traditionally it has been considered that the redecoration and reinstatement covenants of the lease were not subject to the Section 18(1) cap. However, in the case of *Latimer v Carney* [2006], Lady Justice Arden questioned this approach and concluded that decoration may, in certain circumstances, fall within the scope of Section 18(1).

Reinstatement and redecoration are also governed by common law principles under which the landlord's entitlement is to be reimbursed for its financial loss. For this reason an assessment of the landlord's loss is commonly described as a 'diminution valuation'. It will be relevant, however, in assessing the loss, whether the landlord truly intends to reinstate tenant's alterations, for example, a mezzanine floor in a warehouse.

In practical terms the operation of Section 18(1) requires a comparison to be made between two hypothetical market valuations of the premises as demised by the lease. These are as follows:

- Valuation A: assuming compliance;
- Valuation B: assuming actual condition (non-compliance).

The difference between Valuation A and Valuation B assesses the statutory cap (Section 18.1 cap) on the amount of damages payable.

It has been held that even where Valuation A is a negative figure and Valuation B is a greater negative figure then the difference, amounting to the landlord's loss, is still payable (see *Shortlands Investments Ltd v Cargill Plc* [1995]).

The second limb of Section 18(1) deals with what is often referred to as 'supersession', and provides:

> *And in particular no damage shall be recovered for a breach of any such covenant or agreement to leave or put premises in repair at the termination of a lease, if it is shown that the premises, in whatever state of repair they might be, would at or shortly after the termination of the tenancy have been or be pulled down, or such structural alterations made therein as would render valueless the repairs covered by the covenant or agreement.*

> *©Crown copyright material is reproduced under the Open Government Licence v1.0 for public sector information: www.nationalarchives.gov.uk/doc/ open-government-licence/*

The second limb of Section 18(1) would operate, for example, where it could be demonstrated that the landlord of an unmodernised 1960s office building intended, as at the expiry of the lease, to carry out refurbishment works to install a suspended ceiling, a raised floor, and air conditioning.

In such a case, the landlord's proposals could significantly impact upon any remedial repairs required to the interior of the building, rendering valueless the benefit of the required repairs. As a consequence, works that are deemed to be rendered valueless should no longer form part of the claim. However, the claim for those repairs unaffected by the landlord's alterations would remain valid.

The critical date upon which to establish the landlord's intention is the date at the expiry of the lease (see *Salisbury v Gilmore* [1948], *Cunliffe v Goodman* [1950], and, more recently, *PGF II SA v Royal & Sun Alliance* [2010].

The fact that the landlord is contemplating a number of options, including a potential refurbishment at the expiry of the lease, is not a sufficient basis itself to render the value of the repairs nugatory. A clear and fixed intention must be demonstrated.

The Property Litigation Association (PLA) Dilapidations Protocol and RICS Dilapidations guidance note require the surveyor preparing a claim to have regard to the common law principles of loss and Section 18(1) and, as a consequence, both landlords and tenants should obtain advice as to the likely impact of Section 18(1) on their claim at an early stage. A diminution valuation itself may not, however, require preparation until later in the process, depending upon the intended course of action.

3.3 Neighbourly matters

Rights to light
Paul Lovelock

'Rights to light' are private property rights that benefit buildings, both residential and commercial. Not all buildings have them.

The subject of rights to light usually concerns the assessment of whether proposed obstructions (for example, new developments) are likely to interfere materially with neighbours' easements of light. Interpretation of whether a material rights-to-light issue is likely to arise requires knowledge of the law, particularly easements and nuisance, and an understanding of the technical measurements of skylight entering a room. It is then possible to assess the risk of injunction and, where appropriate, the likely level of damages that could be awarded. It is also possible to determine how to modify a proposed scheme in order to reduce or overcome potential problems.

Legal background

Rights-to-light problems bring together two distinct but different areas of English law, namely private nuisance (a subdivision of the law of torts) and easements (a subdivision of land law).

Private nuisance

The tort of private nuisance, like the tort of public nuisance, regulates activities affecting individual rights in or rights over real property (land). A private nuisance may be defined as an unreasonable interference with a person's use or enjoyment of land itself, or some right over or in connection with land (i.e. a right to light).

The law of nuisance tries to balance the legitimate activities of neighbours – a give-and-take approach. Interference with a right to light must be objectively unreasonable in its extent and severity if it is to be sufficient to constitute a nuisance in the eyes of the courts. Only when the courts are satisfied that the interference is unreasonable will they remedy the situation by awarding an injunction and/or damages. It should be appreciated that levels of natural light can often be interfered with to a marginal extent, and this will not necessarily constitute an infringement of a proprietary right that will be recognised as a nuisance.

Easements

An 'easement' may be defined as a right annexed to land to use or to restrict use of neighbouring land in some way. For an easement to be valid, four essential characteristics must be satisfied:

- There must be a dominant tenement and a servient tenement.
- The right must accommodate (benefit) the dominant tenement.
- The dominant tenement and the servient tenement must be owned or occupied by different persons.
- The right concerned must be capable of forming the subject matter of a grant.

Rights to light have been recognised as a valid easement for centuries. The general rights-to-light principles set out below have been distilled from the large body of case law that exists.

A 'right to light' can be defined generally as a negative easement providing a right for a building to receive sufficient natural light through a defined aperture (usually a window), over the land of another, in perpetuity or for a term of years.

Nature of a right to light

A right to light is not personal – it runs with property/buildings. It benefits the dominant tenement and burdens the servient tenement. A right to light is for 'sufficient' natural light only, and this is taken to mean enough light, 'according to the ordinary notions of mankind' for:

- comfortable use and enjoyment of a dwelling house; or
- beneficial use of and occupation of a warehouse, shop, or other place (office, for example).

The test for 'sufficiency' is whether or not the dominant tenement will be left with enough light according to the ordinary requirements of mankind. Sufficiency is not based on the measure of light lost. (See *Colls v Home & Colonial Stores* [1904] AC 179.)

Actionable injury and measurement of light

No specific rule has been developed by the courts to define exactly when a reduction in natural light becomes actionable. The test for injury is uncertain but flexible. The court will have regard, in all cases, to the specific facts and circumstances, and it will usually hear objective technical evidence from a rights-to-light expert, as well as more subjective evidence from the injured party.

A form of technical evidence has evolved that entails analysis of the amount of the notional sky dome that can be seen from a series of points in an affected room at table level. At any given point on the working plane, there is a minimum amount of sky area below which the level of daylight at that point will be inadequate. 'Adequacy' is considered to be just enough for undertaking work that requires visual discrimination, such as reading, drawing, or sewing. In technical terms this is one lumen or 1/500th of a standard uniform dome of overcast sky in December (i.e. 0.2% sky factor).

It is possible to plot the 0.2% sky factor contour in the subject room and measure the area of the room that will receive more than adequate light both before and after development. Today, leading rights-to-light experts tend to do this with the aid of 3D computer modelling and specialist software, which is more accurate and efficient than the laborious manual Waldram method.

Having measured the area that will be adequately illuminated, it is possible to assess whether an actionable injury will arise. Very generally, for day-to-day practical purposes, light specialists have adopted the general conventions that:

- A commercial property may be considered actionably damaged when less than 50 per cent of an office floor area is lit to the critical 1 lumen (0.2% sky factor) level, i.e. the so-called 50/50 rule.
- A residential/domestic property should be considered actionably damaged when less than 55% of a room area is lit to the critical 1 lumen (0.2% sky factor) level.

It must be understood that the percentages mentioned above are not strongly founded in specific legal authority, although courts will make reference to previously decided cases where the facts are similar. The courts regard the so-called '50/50 rule' as a 'convenient rule of thumb'.

Acquisition of a right to light

A right to light may be created by:

- expressed grant or reservation (sometimes encountered);
- implied grant or reservation (rarely encountered); or
- prescription (very common and often called 'ancient lights').

'Prescription' means the procuring of a right on the basis of a long-established custom and three methods of prescription exist, namely:

- time immemorial (right enjoyed since before 1189);
- doctrine of lost modern grant (right enjoyed continuously for minimum 20 years); and
- Prescription Act 1832: sections 3 and 4 (right enjoyed continuously for 20 years).

Note: In the city of London, because of the 'Custom of London', the acquisition of a right by the doctrine of lost modern grant is not available – see *Bowring Services Ltd v Scottish Widows* [1995] 16 EG 206.

Defeating a right to light

Prescription through the doctrines of lost modern grant or time immemorial can be defeated if it can be shown that the easement has not been enjoyed, 'as of right', i.e. through force, secrecy, or with permission.

Statutory prescription, under the Prescription Act 1832, may be defeated if the servient party can show that:

- at some time within the last 19 years, it has prevented the entry of natural daylight through the subject apertures by erecting an opaque physical obstruction for a continuous period of at least 1 year;
- at some time in the last 19 years, it registered a light obstruction notice under the Rights of Light Act 1959 for 1 whole year (see below); or
- the right has been enjoyed under some consent or agreement expressly given for that purpose by deed or in writing.

Defending a right to light

A prescriptive right to light may be lost if it is not defended in the face of development. A neighbour who acquiesces in or submits to an interruption of light for 1 year or more will lose a claim to a prescriptive right under Section 4 of the Prescription Act 1832. (See *Dance v Triplow and Another* [1992] 17 EG 103.)

Successful defence of a prescriptive right to light relies much on eternal vigilance and prompt protestation in writing to the obstructor. The protestations should also be repeated at regular intervals. Ultimately legal proceedings will need to be brought against a developer who ignores the objections, and the timing of this may considerably affect the chances of obtaining an injunction.

Remedies

The current legal system permits the awarding of:

- prohibitory or mandatory injunctions,
- common law damages, and/or
- declarations.

Injunctions

The courts are reluctant to sanction a wrongdoing by a servient party in allowing that party to purchase its neighbour's rights. Generally, an injunction is regarded as the normal remedy, with damages the exception. The Court of Appeal has recently reminded all of this principle in their decision in *Regan v Paul Properties Ltd and Others* [2006] EWCA Civ 1319.

The court may award damages in lieu of an injunction if all the four requirements below can be answered in the affirmative:

- Is the injury small?
- Would a small money payment be an adequate remedy?
- Would it be oppressive to the defendant to grant an injunction?
- Is the injury one that can be estimated in money terms?

These requirements are derived from *Shelfer v City of London Electric Lighting Co* [1895] 1 Ch 287 31, however, the courts have departed from a strict application of the test and it is not necessarily applicable to every rights to light situation.

The conduct of parties will also have a significant bearing on the matter. If a developer ignores the communicated protestations from its neighbour and continues developing regardless then once a nuisance is proved an injunction should be the default remedy.

Damages (compensation)

Compensation for injury to a right to light may be assessed using one of two methods:

- the traditional valuation approach; or
- the share of developer's profit approach.

Compensation/damages based on a share of the developer's profit is the method more likely to be used since the case of *Tamares (Vincent Square) Ltd v Fairpoint Properties (Vincent Square) Ltd* [2007] All ER(D) 1034 (assessment of damages). This method will often derive a significantly larger sum in compensation than will the traditional valuation approach. In *Tamares*, the court set out eight principles, derived from previous cases, to give guidance on how damages should be assessed.

Whichever method is adopted, the valuation approach is based on a freeholder in possession. With a traditional valuation approach, a 'base book value' is calculated for the light loss, which may then be enhanced by a multiplier of up to three or four times, having regard to the case of *Carr Saunders v Dick McNeil Associates Ltd and Others* [1986] 1 WLR 992. Any compensation will be appropriately apportioned between the various interests in the dominant tenement. For leases, this is generally dependent upon the number of years remaining until the next rent review. For residential property, however, the assessment of compensation is more subjective.

Compensation based on a share of the developer's profit requires an (often complicated) assessment to be made of the amount of profit that a developer will be expected to realise if construction of the infringing part of a proposed development actually occurs. The owner(s) of the injured building will then expect a share of that element of profit, which share could be anything between 5% and 50% depending on the circumstances. This approach essentially considers the possibility of sharing the profit that a developer will make from floor space that could not be built if the adjoining owner obtained an injunction. There is no consistent guidance from the courts indicating when profit-based compensation is to be preferred over the traditional valuation approach.

Declarations

Parties seeking certainty in respect of an existing dispute situation can apply to a court of law for determination of an issue by declaration. The declaration will determine the issue in dispute and is effectively a judgement.

Dos and don'ts

- Do establish whether surrounding properties enjoy rights to light, including other tenanted parts of the client's property, and identify all parties with an interest.
- Do establish whether development proposals are likely to leave the surrounding buildings with inadequate light.
- Do take specialist advice from a rights to light consultant at an early stage.
- Do obtain copies of all leases, deeds, transfers, restrictive covenants, and so on that could have a bearing on the legal position.
- Don't be fooled by buildings that look less than 20 years old or have blocked-up windows; case law is complex and there could still be a right to light.
- Do consider whether the Crown has ever had an interest in the development site; the surrounding buildings may not be entitled to rights to light.
- Do consider whether the development site has ever been acquired or appropriated by the local authority for planning purposes under Section 237 of the Town and Country Planning Act 1990, as this affects the potential for injunctions. Local authorities are able to override

rights to light that would otherwise hinder a development by using powers conferred on them by Section 237. The development land concerned has to have been 'appropriated for planning purposes' and the development must be executed in accordance with a planning permission. An injury caused by the proposed development must be addressed by paying the injured party compensation but the sums concerned are calculated on compulsory purchase principles and not the common law principles described above. (Other legislation relating to development by government-funded agencies, for example, the Housing and Regeneration Act 2008, also exists and contains provisions that are similar to Section 237.) Use of Section 237 powers has become more prominent for major landmark developments since the case of *HKRUK II (CHC) Ltd v Heaney* [2010] 44 EG 126.

- Do obtain copies (if available) showing the massing and profile of the existing building on the development site, and the proposed building or extension, and also up-to-date floor layout plans for the surrounding buildings.
- Do establish the extent of a right (i.e. the number, size, and location of apertures).
- Do consider whether transferred or 'incorporated' rights to light exist.
- Do seek the advice of lawyers if the legal position is complicated beyond your experience by any controlling deeds or similar documents.
- Do ensure that your client understands that the law relating to rights to light, and the valuation techniques, are not an exact science.
- Do remember that recent case law has reminded all that the primary remedy of interference with a right to light is an injunction – not damages.
- Do remember that you cannot rely on a neighbour, particularly a residential owner, settling for compensation.
- Do explain to your client that there are no general statutory procedures for dealing with rights to light issues and

that there are no prescribed periods and deadlines during or by which parties are obliged to settle issues.

- Don't allow a 'dominant' party to be pressured into early agreement of compensation.

Light obstruction notices

Light obstruction notices can be used to either defeat an existing right to light that has been acquired by statutory prescription or prevent such a right from being acquired. They are a useful tool for preserving the development potential of a site and can also be used to 'flush out' potential claims. Under Section 2(1) of the Rights of Light Act 1959, a light obstruction notice can be registered with the local authority as a local land charge for a period of 1 year.

Impending changes to the law and procedures relating to rights to light

Rights to light are often seen as one of the main obstacles to development in the UK, along with planning rules and regulations. Those benefiting from rights to light see them as valuable to their use and enjoyment of property. There is, however, pressure from developers for reformation of the law relating to rights to light; the Law Commission has considered and reported the subject.

The Law Commission has thoroughly reviewed the law relating to rights to light and issued its final report (LAW COM No.356) to Parliament in December 2014.

The report recommends:

- A new statutory test of proportionality specific to rights to light. The awarding of an injunction may be seen as disproportionate remedy and unnecessarily aggressive.

- Creation of a statutory notice called a 'Notice of Proposed Obstruction'. This will allow a developer to give neighbours notice that he expects to obstruct the light to their property. The recipient neighbours will then be required to commence proceedings to claim an injunction within 8 months. Should they fail to do so, the right to claim an injunction will be lost, and only a claim for damages will survive.

Enactment of the recommendations made by the Commission will ultimately depend on available parliamentary time and political priority.

References
- *Rights of light: Practical guidance for chartered surveyors in England and Wales* (1st edition), RICS guidance note, 2010.
- *Rights of light: The modern law* (3rd edition), S. Bickford-Smith and A. Francis, Jordan Publishing Ltd, 2015.
- *Anstey's rights of light* (4th edition), J. Anstey and L. Harris, RICS Books, 2006.

Further information
The two main statutes that are relevant to rights to light are:

- the Prescription Act 1832; and
- the Rights of Light Act 1959.

There exists a large body of rights to light case law. The selection of case reports listed below includes the more recent and more important decisions.

Allen and Another v Greenwood and Another [1975] 1 All ER 819 6, 35-6.
Bowring Services Ltd v Scottish Widows Fund & Life Assurance Society [1995] 16 EG 206.

Carr Saunders v Dick McNeil Associates Ltd and Others [1986] 1 WLR 992 37, 43.

CGIS City Plaza Shares 1 Ltd and another v Britel Fund Trustees Ltd [2012] EWHC 1594 (Ch)

Charles Semon & Co v Bradford Corporation [1922] 2 Ch 737.

Colls v Home & Colonial Stores [1904] AC 179 4, 9, 10, 29, 31-2, 34.

Colls v Laugher [1894] 3 Ch 659.

Coventry and Others v Lawrence and Another [2014] UKSC 46.

Dance v Triplow and Another [1992] 17 EG 103.

Deakins v Hookings [1994] 14 EG 133.

Ecclesiastical Commissioners for England v Kino [1880] 14 ChD 213 40.

Fishenden v Higgs and Hill Ltd [1935] 153 LT 128 33.

Forsyth-Grant v Allen and Another [2008] EWCA Civ 505.

HKRUK II (CHC) Ltd v Heaney [2010] 44 EG 126.

Lyme Valley Squash Club Ltd v Newcastle under Lyme Borough Council and Another [1985] 2 All ER 405 28–9.

Marine and General Mutual Life Assurance Society v St James Real Estate Co Ltd [1991] 2 EGLR 178.

Midtown Ltd v City of London Real Property Co Ltd [2005] 14 EG 130.

Ough v King [1967] 3 All ER 859 34.

Price v Hilditch [1930] I Ch 500 5, 37, 43.

Pugh and Another v Howels and Another [1984] 48 PCR 298 36–7.

Regan v Paul Properties Ltd and Others [2006] EWCA Civ 1319.

RHJ Ltd v (1) FT Patten (Holdings) Ltd and (2) FT Patten Properties (Liverpool) Ltd [2007] EWHC 1655 (Ch).

Salvage Wharf Ltd and Another v GS Brough Ltd [2009] EWCA Civ 21.

Scott v Pape [1886] 31 ChD 554 6, 80.

Sheffield Masonic Hall Co v Sheffield Corporation [1932] 2 Ch 17 43.

Shelfer v City of London Electric Lighting Co [1895] 1 Ch 287 31.

Tamares (Vincent Square) Ltd v Fairpoint Properties (Vincent Square) Ltd [2006] EG41 226 (injunction v damages ruling).

Tamares (Vincent Square) Ltd v Fairpoint Properties (Vincent Square) Ltd [2007] ALL ER(D) 1034 (assessment of damages).

Wheeldon v Burrows [1879] 12 ChD 31 78.

Wrotham Park Estate Co v Parkside Homes Ltd [1973] ChD 321.

Daylight and sunlight amenity
Paul Lovelock

As greater emphasis is placed on environmental issues, local planning authorities are now regularly concerning themselves with the effect of developments on the daylight and sunlight enjoyed by neighbouring properties. A local authority's planning policy and strategy documents should always be consulted, as they will give an indication of what is expected in this regard. Also bear in mind supplementary planning guidance issued by the local authority.

The Building Research Establishment (BRE) published the second edition of Report 209 in October 2011 titled *Site layout planning for daylight and sunlight: A guide to good practice*. It is intended to give guidance on, among other things:

- Assessing whether a proposed scheme will enjoy adequate levels of sunlight and daylight amenity
- Assessing the impact that a proposed scheme will have on the levels of daylight and sunlight amenity enjoyed internally and externally by other neighbouring buildings and open areas; it sets out various assessments that can

be undertaken to establish if a noticeable reduction in amenity is likely to result from a scheme

The BRE Report is not intended to be mandatory. However, it is not uncommon for planning authorities to require a developer to submit a daylighting and sunlighting study in support of a planning application, and some may expect strict compliance with the recommendations given in the BRE Report guidelines.

It is important to note that even if a scheme is granted planning consent, this will not override the private rights of individuals, and therefore, rights to light may still be an issue. (See *Brewer and another v Secretary of State for the Environment and others* [1988] 2 PLR 13.)

The BRE Report advocates different assessment methodologies for daylight and sunlight.

Daylight

The BRE Report states that if any part of a new building or extension, measured in a vertical section perpendicular to a main window wall of an existing building, from the centre of the lowest window, subtends an angle of more than 25 degrees to the horizontal, then the diffuse daylighting of the existing building may be adversely affected. This will be the case if either:

- the vertical sky component measured, at the centre of an existing main window, is less than 27 per cent, and less than 0.8 times its former value; or
- the area of the working plane in a room which can receive direct skylight is reduced to less than 0.8 times its former value.

In such circumstances, the occupants of the existing building will notice the reduction in the amount of light from the skylight, and more of the room will appear poorly lit. The assessment criteria above require the calculation of both vertical sky components for windows and no sky lines for rooms.

Another method of assessing the interior daylight levels within a room, called the 'Average Daylight Factor' (ADF), is described in British Standard 8206-Part 2: 2008. There are recommended minimum ADF values for dwellings, where supplementary electric lighting is to be used in rooms, namely 1% for bedrooms, 1.5% for living rooms, and 2% for kitchens. The assessment of Average Daylight Factor is not however a central feature of the methodology set out in BRE Report 209. ADF values do not form any part of daylight and sunlight amenity assessments in BRE Report 209, however.

Sunlight

The BRE Report advises that new development should take care to safeguard access to sunlight for existing dwellings and any non-domestic buildings where there is a particular requirement for sunlight. Sunlight is quantified for windows (rather than rooms) using a measure called Annual Probable Sunlight Hours. In the northern hemisphere only those windows facing with 90 degrees of due south will benefit from sunlight – windows facing within 90 degrees of due north need not be assessed. Sunlight is not the same in amenity terms as passive solar energy. The latter is dealt with by the BRE Report, but is not normally relevant to planning applications.

The BRE Report states that if a living room of an existing dwelling has a main window facing within 90 degrees of due south, and any part of a new development subtends an angle of more than 25 degrees to the horizontal, measured from the centre of the window in a vertical section, perpendicular to the window, then the sunlighting of the existing dwelling may be adversely affected. This will be the case if the centre of the window:

- receives less than 25% of annual probable sunlight hours, and/or less than 5% of annual probable sunlight hours during the winter months between 21 September and 21 March
- receives less than 0.8 times its former sunlight hours in either the annual or the winter period, and
- has a reduction in sunlight received over the whole year greater than 4% of the annual probable sunlight hours enjoyed by that window.

Overshadowing

BRE Report 209 also provides guidance concerning overshadowing created by a proposed scheme. It notes that sunlight to open spaces is "valuable for a number of reasons", namely to:

- provide attractive sunlit views (all year);
- make outdoor activities like sitting out and children's play more pleasant (mainly during the warmer months);
- encourage plant growth (mainly in spring and summer);
- dry out the ground, reducing moss and slime (mainly in the colder months);
- melt frost, ice, and snow (in winter); and
- dry clothes (all year).

The BRE Report suggests that the availability of sunlight should be checked for all open spaces where it will be required. It goes on to state that this would normally include gardens, allotments, parks, and playing fields, children's playgrounds, outdoor swimming pools, paddling pools, public sitting-out areas, and focal points for views (such as a group of monuments or fountains).

BRE Report 209 gives guidance concerning the degree of change in overshadowing to open areas, resulting from an adjoining development, which is thought to be permissible before the change becomes noticeable to users of the space. Under the report, at least 50% of a garden or amenity area should receive 2 hours sunlight on 21 March. If, due to development work, the area is reduced to 0.8% of its former value, then the loss of sunlight is likely to be noticeable.

Analysis

The BRE Report sets out the methods by which the effect of a development on existing neighbouring buildings and open spaces may be assessed. If the local planning authority requires an assessment to be undertaken and submitted in support of a planning application, it is usual for the analysis to follow the methodologies in BRE Report 209.

The target values recommended in BRE Report 209 are quite demanding and can be difficult to achieve, particularly in dense urban environments. Care should be taken when applying the guidance and interpreting the results of analyses.

Further information

BRE Report 209, *Site layout planning for daylight and sunlight: A guide to good practice* (2nd edition, 2011), by Paul. J. Littlefair. BRE.

BS 8206-2: 2008, *Lighting for buildings: Code of practice for daylighting.* BSI.

Brewer and another v Secretary of State for the Environment and others [1988] 2 PLR 13.

Planning appeal by St George Central London, decided 20 April 2004.

Malster v Ipswich Borough Council [2001] EWHC ADMIN 711.

Planning appeal by West End Green (Properties) Ltd, decided 10 October 2005.

Planning appeal by Inner Circle Ltd, decided 16 March 2007.

Party wall procedure
Paul Lovelock

The Party Wall etc. Act 1996 came into force throughout England and Wales on 1 July 1997. All previous local enactments, including the London Building Acts (Amendment) Act 1939 Part VI and the Bristol Improvement Act 1847, have now been repealed.

The general principle of the Act is to enable an owner to undertake certain specific works on, or adjacent to, adjoining properties while giving protection to potentially affected neighbours. As suggested by the word 'etc.' in its title, the Act relates not just to party walls. Its purpose is to facilitate the execution of work, not to prevent it. Today's legislation is the latest incarnation of legislation that came into being, in a form recognizable today, as a result of the Great Fire of London in 1666.

The parties

The owner of the property where the work is to be undertaken is the 'building owner'. The owner of the adjoining property is the 'adjoining owner'. There can be many adjoining owners, including freeholder, leaseholder, and anyone with an interest greater than from year to year. Under the Act, the word 'owner' can also include people with a contract to purchase or an agreement for lease; this arrangement allows a prospective owner to serve notice, and even commence work, before completion of the contract.

Party structures

A party wall is one that stands on the land of two owners, by more than its footings, or one which separates buildings of different owners. In the first case, the whole wall is a party wall, whereas in the second, it is only a party wall for the extent to which the two buildings are using it and the whole of the rest of the wall belongs to the person on whose land it stands. Other types of party structures are party fence walls (i.e. shared garden walls) and party floors (e.g. separating different flats).

In the case of a party wall, each owner owns the part of the wall that stands on their own land, but also has rights over the remainder of the wall. The Act allows owners to treat the whole of the party wall as if it were their own, and debars them from dealing with their half on its own without informing their neighbours.

Before exercising any of the rights bestowed upon them, owners must follow the procedures set down in the Party Wall etc. Act 1996. If an owner wishes to underpin, raise, cut into, thicken or demolish and rebuild a party wall, that owner must give notice of his or her intention to do so. In the event that the adjoining owner disagrees, each party must appoint a surveyor and a formal agreement known as a party wall award must be entered into.

An adjoining owner may respond to a party-structure notice by serving a counter notice requiring certain speci- fied works to be undertaken to the party wall for his or her future purposes. However the notice must be served within 1 month of the original notice, and the adjoining owner will be responsible for the cost of the additional work.

Excavations adjacent to structures

The Party Wall etc. Act 1996 requires notice to be served on adjoining owners of certain intended excavations within 3 metres or 6 metres of adjacent buildings or structures. The excavations could be for any purpose, not just for a building or its foundations.

Refer to the following diagrams for details of these notifiable excavations.

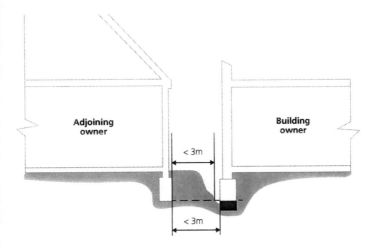

Figure 3.2 Three-metre notice

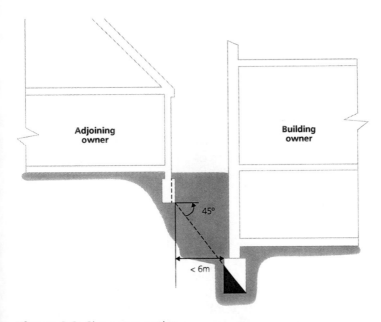

Figure 3.3 Six-metre notice

New walls at boundaries

Where it is proposed to build at the 'line of junction' (i.e. the boundary), notice may need to be given to the adjoining owners. This will be the case where the boundary is not already built upon, or is built upon only to the extent of a boundary wall (not being a party fence wall or external wall of a building). In return, a right to construct projecting mass concrete footings work may be claimed.

General rights and obligations

The Party Wall etc. Act 1996 grants various general rights and obligations in relation to work undertaken in pursuance of the Act. The key ones are:

- obligation to serve notice on adjoining owners before undertaking the work;
- obligation to execute the work in compliance with other statutory requirements and in accordance with details agreed either by the parties or by appointed party wall surveyors in an award;
- ability to enter and remain on adjoining owners' land so far as is reasonably necessary to facilitate the work;
- duty not to cause unnecessary inconvenience;
- duty to compensate adjoining owners for loss or damage arising as a consequence of the work; and
- obligation to make good damage caused by work to party structures.

Agreeing the works

The building owner and adjoining owner may agree the details of the works between them. Where disagreements occur, either deemed or actual, the Act provides a mechanism for resolving them through the appointment of party wall surveyors. The appointed surveyors have a duty to settle the matter by making an award and they must act expediently and impartially, having regard to the interests and rights of both parties and the extent of the statutory powers, which are not without limit.

Entry to adjoining land

Section 8 of the Act confers certain rights of entry on the developing party. There are differing views on whether a building owner may enter adjoining land for all construction operations covered by the Act, or only where the Act expressly grants a right to do work, such as raising a party wall. The Act is ambiguous, which may potentially create difficulties for building owners who wish to build a new wall wholly on the line of junction and who need to erect scaffolding on adjoining land. For such work it is important to establish the views of the surveyors and adjoining owners concerned at an early stage.

Dos and don'ts

- Do consider whether proposed excavations and building works require the serving of notices under the Act.
- Do allow adequate time in the project programme for identifying and resolving all party wall issues. Time periods stated in the Act are statutory minimums, and often longer periods should be allowed.
- Do establish which party will act as the 'building owner' and verify his or her legal interest in the property.
- Do remember that if the building owner parts with his or her interest in the property part-way through proceedings, matters will have to start afresh.
- Do remember that the interests of all relevant adjoining owners will have to be identified with certainty and each owner notified separately.
- Do remember that each adjoining owner has a right to disagree with the proposals and to appoint a surveyor.

- Don't let unreasonable or unresponsive adjoining owners or surveyors hinder progress: use the mechanisms in the Act to force matters along.
- Do make allowance within the cost plan for the reasonable fees of the adjoining owners' surveyors and any necessary subconsultants (typically structural engineers).
- Do ensure that the design team and contractors cooperate and produce all necessary drawings, method statements, calculations, etc. in good time.
- Do ensure that the proposed time and manner of carrying out the work is reasonable and that no unnecessary inconvenience will be caused.
- Do consider making provision within tender documents for restrictions on working hours and/or methods of working for noisy elements of work falling within the scope of the Act.
- Do consider whether access will be required onto adjoining land to build a wall at the line of junction and bear in mind that differing views exist as to whether a right of entry exists for such work.
- Do ensure that the procedures required by the Act are followed meticulously; otherwise, notices and awards could be invalidated.
- Do remember that the party wall surveyors are administering the Act impartially and not representing clients.
- Do ensure that awards are agreed prior to starting the relevant work.
- Don't confuse common-law and rights-to-light matters as being part of party wall procedures.
- Do establish a line of communication between adjoining owners and contractors for dealing with day-to-day issues of noise, dust, and so on.
- Do ensure that any variations in the agreed works are agreed between the parties, or by their appointed surveyors.
- Finally, do ensure that all parties fulfil their obligations.

Further information

- *Party wall legislation and procedure* (5th edition), RICS guidance note, 2002.
- *The Party Wall Act explained – A commentary on The Party Wall etc. Act 1996* (The Green Book), Pyramus and Thisbe Club, 2nd edition, 1997.
- *Party walls – Law and practice* (3rd edition), S. Bickford-Smith, C. Sydenham, and A. Redler, Jordans, 2009.
- *A practical manual for party wall surveyors*, J. Anstey, RICS Books, 2000.
- *Anstey's party walls and what to do with them* (6th edition), G. North, RICS Books, 2005.
- *An introduction to the Party Wall etc. Act 1996*, J. Anstey and V. Vegoda, Lark Productions, 1997.
- *The Party Wall Casebook*, P. Chynoweth, Blackwell Science, 2008.
- *Party Wall etc. Act 1996*, audio cassette, Owlion.
- *Party walls – Best practice roadshow*, audio CD, Owlion, 2002.
- *Party Wall etc. Act, 1996: Explanatory booklet*, Department for Communities and Local Government.
- *Practical neighbour law handbook*, A. Redler, RICS Books, 2006.
- *Boundaries: Procedures for boundary identification, demarcation and dispute resolution in England and Wales* (2nd edition), RICS guidance note, 2009.

Useful websites

Party Wall etc. Act 1996 – Office of Public Section Information: www.legislation.gov.uk/ukpga/1996/40/contents
Royal Institution of Chartered Surveyors: www.rics.org/partywallsguide
Pyramus and Thisbe Club: www.partywalls.org.uk
Department for Communities and Local Government (DCLG): www.gov.uk/government/organisations/department-for-communities-and-local-government

Access to adjoining property
Paul Lovelock

The safe and economic execution of construction work on, or close to, site boundaries frequently requires access to adjoining land for plant and/or materials and/or operatives.

Rights to access land belonging to adjoining owners for specific construction works exist in the form of:

- the Access to Neighbouring Land Act 1992
- the Party Wall etc. Act 1996
- compulsory purchase legislation

This legislation does not provide access rights for every conceivable construction activity in every conceivable circumstance, and, when access rights do not exist, express consent from adjoining owners has to be sought for such access. Express consent is often given in the form of a licence.

Access to Neighbouring Land Act 1992

Under the Access to Neighbouring Land Act 1992, a court may grant an order for access to land where such access is required to enable the execution of "basic preservation works" and where access has been refused by the neighbour. Such works include:

- maintenance, repair, or renewal of any part of a building or structure;
- clearance, repair, or renewal of drains, sewers, pipes, and cables;
- cutting back or felling trees and hedges in certain circumstances; and
- filling in or clearing any ditch.

The Act assists in facilitating the repair and maintenance of buildings and structures, its purpose is not to assist in redevelopment, alteration or extension of buildings and structures. The provisions of the Act are administered by solicitors and effectively involve county court litigation. The Act contains provisions for the preparation of schedules of condition and for overseeing of the work by surveyors. Surveyors can also be called upon to provide evidence where claims for damage under the Act are made.

The Order for access made by the court may require the payment of consideration, having regard to the likely financial advantage to the applicant (developing party) and the degree of inconvenience, except where the works are to residential land.

Often the knowledge that rights of access exist under the Act will prompt neighbours to be accommodating, but, even then, it is still sensible to enter into an informal access agreement in the form of a licence with accompanying schedule of condition.

Further information
This Act is available online at: www.legislation.gov.uk/ukpga/1992/23/contents

3.4 Workplace

Health and safety at work
Paul Winstone

Essential legal duties

The Health and Safety at Work etc. Act 1974 requires all surveyors to ensure, so far as is reasonably practicable, the health and safety of themselves and any other people who may be affected by their work.

In essence, the places where surveyors work must be safe and working practices must be clearly defined, organised, and followed to avoid danger. This requires safety training and the distribution of relevant information, followed up by diligent and regular supervision.

These duties extend to anyone who uses surveyors' professional services.

Those who lease part of their premises to other businesses also may be responsible for them with regard to safety matters.

In addition, all working people, whether employees, managers, partners, or directors (self-employed or not) must behave in a way that does not endanger themselves or others.

Health and safety policy statement

Firms that employ five or more people are obliged by the Act to draw up a health and safety policy statement, which should be kept up to date, with any significant revisions being notified to employees.

The Health and Safety Executive (HSE) in *Writing your health and safety policy statement: How to prepare a safety statement for smaller businesses* describes what the document should say and gives a useful pro forma.

It is suggested that employees, especially when taking a new job, should satisfy themselves that they have seen and understood the company's policy statement and have been made familiar with safe working practices. If they consider that the Health and Safety at Work Act is not being followed, they ought to say so, and if necessary ask advice from their local HSE office (listed in the telephone directory).

Practical procedures

Surveyors should identify the hazards they may encounter in practice, carry out a risk assessment, and plan accordingly. The way they proceed will depend upon the working environment: when surveying old and derelict buildings, for example, there may be holes in floors, parts of the structure may be unstable, or there may be health hazards, and particular care should be taken to protect against personal attack.

The Control of Substances Hazardous to Health Regulations 2002 (as amended)

The purpose of the Control of Substances Hazardous to Health Regulations 2002 (as amended in 2004) is to safeguard the health of people using or coming into contact with substances that are hazardous to health.

Substances are classified as being very toxic/toxic, harmful, corrosive, or irritant. Under these Regulations employers are required to evaluate the risk of all products used that may be harmful to the health of their employees and take appropriate measures to prevent or control exposure.

The 'Six Pack' Regulations

The original six sets of Regulations were introduced in 1993. They were wide ranging, and, with some minor exceptions, applied to all places of work, replacing and consolidating various individual acts or regulations applicable to individual industries or sectors of industry, such as the Factories Act, the Offices, Shops and Railway Premises Act, and the Construction Regulations.

Provision and Use of Work Equipment Regulations 1998 (as amended by The Health and Safety [Miscellaneous Amendments] Regulations 2002)

These apply to all workplaces. Basically, all existing and new work equipment, which includes everything hired or purchased second hand, must comply with the Regulations.

Every employer must ensure that all work equipment is so constructed or adapted as to be suitable for the purpose for which it is used or provided. They identify specific hazards that the employer must prevent or adequately control.

Manual Handling Operations Regulations 1992 (as amended) (MHOR)

These require the employer to try to avoid the need for employees to undertake any manual handling operations at work that involve a risk of their being injured. Where this is not reasonably practicable the risk must be assessed and suitable provision made, including equipment, instruction, and training for safe manual handling.

Management of Health and Safety at Work Regulations 1999

These Regulations are of a wide-ranging general nature and overlap with many others. They require the employer to carry out an assessment of the risks of the hazards to which his or her employees are exposed at work or to others arising from or in connection with this work. The employer must instigate appropriate protective or preventive measures, reviewing and amending these as necessary.

The employer must appoint a 'competent person' to provide assistance in respect of these duties. Emergency procedures must be put in force to deal with any serious and imminent danger. Employees must be informed of these measures and suitably trained where required. They are obliged to comply with these instructions and warn of any situation considered to be a serious and immediate danger to health and safety.

Personal Protective Equipment at Work Regulations 1992

Under these Regulations the employer has a duty to provide and maintain suitable personal protective equipment, including adequate instruction and training on its correct use when risks to health and safety cannot be avoided by other means. Employees have a duty to make full and proper use of such equipment provided, and to report any loss or obvious defects.

Display Screen Equipment Regulations 1992

The need for the Regulations is primarily the evidence of repetitive strain injury that is reported by keyboard users, the amount of time lost due to other causes of sickness among users and, of course, the European Directive.

These Regulations target full-time users of visual display units, mainly in the banking, insurance, and data-processing sectors, but, given that most medium and larger companies will have in their offices a number of full-time or habitual users, then these Regulations will apply. They will also apply in the office facility of a construction site, if any persons are habitual users of visual display units.

Workplace (Health, Safety and Welfare) Regulations 1992

These apply to all workplaces. The Regulations set out the minimum requirements in respect of the provision and maintenance of the environmental and working conditions of employees. In addition, they impose particular safety requirements on forms of construction or circumstances which are considered to be high risk. The Workplace Regulations do not apply to construction sites (see <u>The Construction (Design and Management) Regulations 2015</u>).

The following issues are addressed by the Workplace Regulations:

- ventilation;
- temperature;
- lighting (see below);
- cleanliness and waste materials;
- room dimensions and space;
- maintenance of the workplace;
- floors and traffic routes;
- ('glazing') transparent or translucent doors, gates, walls, and windows – protection against breakage;
- windows, doors, and gates – cleaning and safety;
- escalators and moving walkways;
- sanitary conveniences and washing facilities;
- drinking water;
- accommodation for work clothing and changing facilities;

- facilities for rest and eating facilities; and
- working at height.

Workplace regulations – glazing
Paul Winstone

Regulation 14 of The Workplace (Health, Safety and Welfare) Regulations 1992 requires that glazed doors (and gates) be fitted with safety materials where any part of the glazing is below shoulder height. This requirement applies to glazing in the panels at the sides of the doors (and gates) because these areas are often struck or pushed when mistaken for part of the door.

The requirements also apply to windows, walls, and partitions where there is glazing below waist height.

In situations where the width of the glass panel exceeds 250mm, then safety materials must be used. Safety materials include:

- polycarbonates, glass blocks, or other materials that are inherently robust; or
- glass that will break safely (i.e. shatters without a chance of sharp edges) or ordinary annealed glass that is of sufficient thickness relative to its area, as outlined in the following table.

Table 3.3

Nominal thickness	Maximum size
8mm	1100 x 1100mm
10mm	2250 x 2250mm
12mm	3000 x 4500mm
15mm	Any size

Therefore, just because annealed glass exists, it does not follow that additional protection is required.

Glazing should always comply with British Standard 6262: Part 4: 2005 *Glazing for building: Code of practice* for safety related to human impact.

In circumstances where glazing does not comply, it will be necessary to replace it with safety materials or to provide some physical protection that will ensure that it meets the impact performance levels required by BS 6262, Part 4: 2005.

A cheaper alternative would be to apply safety film to achieve the BS 6262 standard. Manufacturers of film should be consulted to ensure an appropriate grade of material for glass size.

All glass should be suitably marked as being of a safety standard. Where glass is protected by film, labelling should identify this.

Identification

Laminated and toughened glass can be detected using proprietary glass-testing kits, for example, the Merlin laser glass measurement gauge (www.merlinlazer.com/Glass-Testing-Measurement).

Note: ordinary Georgian wired glass does not comply with safety standards, but Georgian wired safety glass does.

Workplace regulations – provision of sanitary facilities
Paul Winstone

The Workplace (Health, Safety and Welfare) Regulations 1992 require that sanitary provision shall be suitable and sufficient for the numbers and types of workers employed.

The Equality Act 2010 means that other factors will also need to be considered, perhaps including the provision of unisex WCs to allow people not to disclose their physical gender.

Designing with transgender in mind may be relatively easily catered for in offices, but in single-sex accommodation in hospitals or educational establishments, this could prove to be more difficult.

The Approved code of practice and guidance L24 gives minimum numbers of facilities:

Table 3.4 Number of toilets and washbasins for mixed use (or women only)

Number of people at work	Number of toilets	Number of washbasins
1–5	1	1
6–25	2	2
26–50	3	3
51–75	4	4
76–100	5	5

Note: The number of people at work shown in Column 1 refers to the maximum number likely to be in the workplace at any one time.

Table 3.5 Toilets used by men only

Number of men at work	Number of toilets	Number of urinals
1–15	1	1
16–30	2	1
31–45	2	2
46–60	3	2
61–75	3	3
76–90	4	3
91–100	4	4

Workplace regulations – working at height
Paul Winstone

Background

Falls from height are the most common cause of fatal injury and the second most common cause of major injury to employees in the workplace. Most falls are the result of poor management rather than equipment failure. Common examples include:

- failure to recognise a problem;
- failure to provide safe systems of work;
- failure to ensure that safe systems of work are followed;
- failure to provide adequate information, instruction, training, or supervision;
- failure to provide safe plant or equipment; and
- failure to use appropriate equipment.

Regulations

The principles of good practice to prevent falls are contained within the Work at Height Regulations 2005. These Regulations came into operation on 6 April 2005 and apply to all work where there is a risk of a fall liable to cause personal injury, even if the fall is at or below ground level.

The Regulations place duties on employers, the self-employed, and anyone who controls the work of others.

Employers must do all that is reasonably practicable to prevent falling by following a hierarchy for managing and selecting appropriate equipment:

- avoid work at height if possible;

- if unavoidable, use work equipment to prevent falls;
- where the risk of fall cannot be eliminated, use equipment or other measures to minimise the distance and consequences of any fall; and
- select collective measures to prevent falls (e.g. guardrails and working platforms) before measures which may only mitigate the distance and consequences of a fall (e.g. nets or airbags) or which may only provide personal protection from a fall (e.g. fall arrest lanyards).

Employees must report any safety hazards and properly use the equipment supplied to them.

Duty holders must ensure that:

- all work is properly planned and organised;
- work activities take account of weather conditions;
- persons involved are trained and competent;
- the place of work is safe;
- equipment is appropriately inspected; and
- risks arising from fragile surfaces and from falling objects are properly controlled.

Useful information

The Health and Safety Executive (HSE) has produced a number of free leaflets on falls from height that can be downloaded from their website at www.hse.gov.uk/ falls. These include general advice, such as *Safe use of ladders and stepladders*, as well as the use of specific equipment such as *Safety in window cleaning using rope access techniques (as updated)*.

Further information
See www.hse.gov.uk

4
Materials, defects, and cladding

4.1 Materials

Deleterious materials
Trevor Rushton

The property and construction industry often refers to the term 'deleterious materials' in the absence of an industry-wide agreed definition of exactly what materials are considered deleterious or harmful. Many materials, if used incorrectly, can perform badly but equally, when used in accordance with their known working parameters, can perform perfectly well. In many respects it is better to be more specific and identify materials according to their propensity to cause harm – either to people or to buildings.

The property market has grown accustomed to a range of materials that it considers to be deleterious. In some cases the description is fair and reasonable (for example, asbestos); in other cases it is unjustified. The presence of deleterious materials in a building may affect its market value and could, in severe cases, result in element failure or affect the health of persons working or living there.

The reaction of investing institutions to these materials depends on a number of factors, and often the presence of a deleterious substance will not prevent a purchase. However, great care must be taken to assess the actual risks or consequences involved so that a proper judgment can be made.

Materials hazardous to health

The more common hazardous materials, and associated risks, are identified in the following table.

Table 4.1

Materials	Common use	Use risk
Lead	When used in water pipes and lead paint (lead roofing materials pose little or no risk).	Risk of contamination of drinking water in lead pipes, or from lead solder used in plumbing joints. Risk of inhalation of lead dust during maintenance of lead-based paint. Risk to children of chewing lead painted surfaces (Pica). Concentration of lead in paint now generally much reduced. Beware of lead content in brass fittings.
Urea Formaldehyde foam	Cavity wall insulation. Some insulation boards but rare in UK.	There is some evidence that UF foam may be a carcinogenic material, although this is not proven. Vapour can cause irritation. Poorly installed insulation can lead to passage of water from outer leaf of brick to inner leaf in cavity wall situation. There are some worries over formaldehyde used as an adhesive in medium density fibreboard and chipboard but this is likely to be a problem only in unventilated areas with large amounts of boarding.
Asbestos (see also *Asbestos*)	Commercial and residential buildings as boarding, sheet cladding, insulation and other uses, particularly in the 1950s, 1960s, and 1970s.	Airborne asbestos fibres may be inhaled and eventually lead to either asbestosis, lung cancer, or mesothemelioma.

Materials	Common use	Use risk
Machine made mineral fibres (MMMFs)	Common in contemporary construction	MMMFs are classed as carcinogenic, but there are certain derogations from that classification and in practice the materials that are most commonly used in buildings are considered (currently) not to cause harm to human health (other than in terms of local irritations and similar transitory reactions). However, certain types of fibre, such as special ceramic fibres, are considered harmful – these are rare in buildings and are more likely to be found in specialised industries. Despite the above, precautions need to be taken when fibres are of respirable size. Generally health trials are based on investigations of laboratory animals; the fact that health risks to humans have not been fully established does not mean that they are 'safe' – it simply means that at present a relationship has not been established.
Formaldehyde	Common in adhesives, furnishings, dry cleaning fluids, and the like	Formaldehyde is a naturally occurring organic compound in widespread use particularly for the manufacture of resin binders in particleboards and mdf. Some off-gassing can occur where there are large quantities of material, but in a well-ventilated area this will dissipate. Has been classified as carcinogenic to humans.

Materials damaging to buildings

The fact that materials have been classed as deleterious does not necessarily make them deleterious in practice – particularly if they are used within their normal operating parameters and in accordance with published guidance.

Those materials which may affect building performance or structure are identified in the following table.

Table 4.2

Materials	Common use	Use risk
Calcium silicate brickwork	Used in lieu of concrete or clay bricks, often as an inner leaf in cavity work. Often cited as deleterious but if used correctly will perform well.	Calcium silicate brickwork shrinks after construction with further movement due to wetting. Construction must provide measures of control to distribute cracking. Concrete bricks may display a similar propensity to shrinkage and again care must be taken in the design of movement joints, etc. Note that concrete bricks perform in ways that are very similar to calcium silicate bricks and yet it is rare to consider them as deleterious.
Calcium chloride concrete additive	Commonly used in in-situ concrete as an accelerator and often added in flake form. Often found in buildings constructed before 1977. May also be present from atmospheric or traffic exposure.	Reduces passivity of concrete in damp conditions. Subsequent risk of corrosion of steel reinforcement. See also *Calcium chloride*.
High Alumina Cement (HAC) (see also *High Alumina Cement concrete (Calcium Aluminate Cement)*)	Mainly used in the manufacture of precast X or I roof or floor beams together with some lintels, sill members, etc. between 1954 and 1974. HAC was first produced commercially in the UK in 1925.	Strength of concrete can decrease significantly, often when high temperatures and/ or high humidity is involved. Defects may be due to faulty manufacture.

Materials	Common use	Use risk
Sea-dredged aggregate not in compliance with BS EN 1260 (previously BS 882)	In-situ concrete or precast concrete	May contain salts such as sodium chloride. If salts are not properly washed out, risk of corrosion reinforcement. Provided the aggregates are properly washed and controlled in accordance with British Standard requirements, the indications are that there are no greater risks involved than with the use of aggregates from inland sources. Risk of inclusion of reactive aggregates that could contribute to ASR, although this is unlikely with most UK-sourced aggregates.
Mundic blocks and Mundic concrete	Concrete blocks and concrete manufactured from quarry shale, commonly found in the West Country	Loss of integrity in damp conditions. Further research required to identify level of risk across the country.
Woodwool slabs (also woodcrete and chipcrete)	Often used as (a) decking to flat roofs, or (b) permanent shuttering	Use in (a) may be considered acceptable providing material is kept dry; and (b) as a permanent shutter may result in grout loss (honeycombing) or voidage of concrete near to or surrounding reinforcement, particularly with ribbed floors. May result in reduced fire resistance, reinforcement corrosion or in extreme cases loss of structural strength. May be repaired by application of sprayed concrete. Condition investigated by cut-out removal of woodwool in many locations.
Brick slips	Typically 1970s and 1980s to conceal flow nibs in cavity walls	Risk of poor adhesion, lack of soft joints can transfer load to slips and cause delamination.

Materials	Common use	Use risk
Toughened glass containing nickel sulphide impurities	All glass contains impurities but the toughening process converts some particles into an unstable state	Nickel sulphide inclusions can result in the spontaneous fracturing of toughened glass. The heat treatment process causes the inclusions to shrink and become unstable. Over a period of months to years, the particles revert to their stable condition which involves reverting to their previous size. Failures after 20 years are not unknown. The risk of failure can be controlled by applying a further 'heat soaking' treatment during manufacture. Because of the risk of failure, toughened glass in overhead situations is best avoided.

Materials that can have harmful effects on buildings

Materials that have not been classed as hazardous or deleterious but which can have harmful effects on a building are identified in the following table.

Table 4.3

Materials	Common use	Use risk
Clinker concrete	Typically late 19th- and early 20th-century construction for fire resisting floors reinforced with steel joists	In damp conditions, produces sulphuric acid from combustion products and unburnt coal in the clinker concrete; the acid has a corrosive effect on steel joists leading to loss of section.
Masonry encased steel	Typically late 19th- and early 20th-century construction.	Corrosion of steel frame due to poor protection against moisture and corrosion. Results in cracking and possible dislodgement of building stone.

Materials	Common use	Use risk
Marble cladding	Late 20th-century construction using thin stone panels as cladding (does not affect ashlar)	Natural characteristics of calcitic and dolomitic marbles lead to anisotropic movement and thermal hysterisis. Bowing and sugaring of marble panels is prevalent leading to eventual failure. Process is irreversible.

Further information

For a fuller description refer to:

- *Investigating hazardous and deleterious building materials*, Rushton, T., RICS Books, 2006. Report 325.
- *Sulfide-related degradation of concrete in Southwest England* ('The Mundic Problem') BRE, 1997.

Asbestos

Paul Winstone

'Asbestos' is the generic term for several mineral silicates occurring naturally in fibrous form. Because of its various useful properties (resistance to heat, acids and alkalis, and good thermal, electrical, or acoustic insulator) it has been extensively used in the construction industry.

Three main types used in the UK were Chrysotile (white), Amosite (brown), and Crocidolite (blue). It was used in a variety of forms varying from boards or corrugated sheets to loose coatings or laggings and, generally, the more friable the material, the greater the asbestos content.

Health risk

Inhalation of its microscopic fibres can constitute a serious health risk and is associated with several terminal diseases.

Not all asbestos, irrespective of its circumstances, constitutes an immediate risk, although the effects of possible future disturbance or deterioration must be considered. Factors to be taken into account are the type, form, friability, condition, and location of the source material.

The (1987) Joint Central and Local Government Working Party on Asbestos concluded that "Asbestos materials which are in good condition and not releasing dust should not be disturbed. ... Materials that are damaged, deteriorating, releasing dust or which are likely to do so should be sealed, enclosed or removed as appropriate. Materials which are left in place should be managed and their condition periodically reassessed. The risk to the health of the public from asbestos materials which are in sound condition and which are undisturbed is very low indeed. Substitute materials should be used where possible provided they perform adequately".

The Health and Safety Executive (HSE) actively discourages the unnecessary removal of sound asbestos materials, and each case should be decided on its own merits following an assessment of the risks arising.

In the past, the people most at risk have been workers in the asbestos industry involved in the importation, storage, manufacture, and installation of materials or components containing this material.

These activities are now banned in the UK and the removal, treatment, or intentional working with asbestos is strictly controlled and generally limited to specialists. The risk however continues for anyone who inadvertently disturbs the asbestos in the course of their routine business, including builders, maintenance workers, electricians, and the like, and the most recent legislation is intended for their protection.

Legislation

The enabling act for asbestos legislation is the Health and Safety at Work etc. Act 1974 and failure to comply with its requirements is a criminal offence.

The principal Regulations that apply to works that could expose persons to the risk of respirable asbestos fibres are the Control of Asbestos Regulations 2012 (CAR 2012) which came into operation on 6 April 2012. In addition, the HSE has produced various Approved Codes of Practice (ACOP) and guidance.

The principal codes of practice are:

L143 – Managing and Working with asbestos
HSG 247 – Asbestos: The Licensed Contractor's Guide

Note: At the time of drafting, the third edition of the RICS Guidance Note *Asbestos and its implications for surveyors and their clients*, published in October 2011, has not yet been amended to reflect the changes implemented by CAR 2012.

CAR 2012 made very few changes and only in respect of exemptions in relation to low-risk work with asbestos. These changes were made in response to the European Commission's opinion that the UK had failed to fully implement Article 3(3) of the corresponding Directive. Article 3(3) deals with exemptions from the application of the regulations and that CAR 2006 had applied these exemptions more widely than the Directive allows.

As well as legislation specifically focused on asbestos, more general health and safety legislation that may also need to be taken into account includes:

- the Management of Health and Safety at Work Regulations 1999; and

- the Construction (Design and Management) Regulations 2007.

For a detailed summary of CAR 2012, see the section *Management of asbestos*.

Application of the regulations

In principle, all of the regulations apply to any type of asbestos work unless the work falls within very specific circumstances as set out in Regulation 3(2). The regulations exempted are:

Table 4.4

Regulation	Title	Details
8	Licensing of work with asbestos	Works to be carried out by a licensed asbestos contractor
9	Notification of work with asbestos	Enforcement authority to be informed of proposed works
18(1)(a)	Asbestos areas	Designation of work areas where employee(s) are liable to be exposed to asbestos and specified precautions are required
22	Health records and medical surveillance	Requirement for periodic medical inspection of employees and retention of records

Regulations 9 (notification of work with asbestos), 18(1)(a) (designated areas), and 22 (health records and medical surveillance) do not apply where:

a. *the exposure to asbestos of employees is sporadic and of low intensity; and*
b. it is clear from the risk assessment that the exposure of any employee to asbestos will not exceed the

control limit [0.1 fibres per cubic centimetre of air averaged over a continuous four-hour period]; and

c. the work involves:

i. *short, non-continuous maintenance activities in which only non-friable materials are handled, or;*

ii. removal without deterioration of non-degraded materials in which the asbestos fibres are firmly linked in a matrix, or;

iii. encapsulation or sealing of asbestos-containing materials which are in good condition; or

iv. air monitoring and control, and the collection and analysis of samples to ascertain whether a specific material contains asbestos.

©Crown Copyright material is reproduced with the permission of the Controller of HMSO and the Queen's Printer for Scotland.

The phrases in italics are explained in the Approved Code of Practice (ACOP) and guidance accompanying the Regulations and should be referred to in detail.

Control limit

This is a specific level of concentration of asbestos in the atmosphere, measured in accordance with the World Health Organisation (WHO) recommended method, or approved equivalent. It is the trigger point for the application of specific regulations and/or controls.

If the risk assessment for the works indicates it is liable to be exceeded, then Regulation 18(1)(2) 'Designated areas' requires the establishment of 'respirator zones', where access is restricted to 'competent' persons and suitable respirators must be worn at all times. In addition, if it is not liable to be exceeded, then this is one of a number of circumstances set out in Regulation 3(2) for which specific regulatory require-ments **may** not necessarily apply.

The current control limit is 0.1 fibres per cubic centimetre of air averaged over a continuous period of 4 hours.

Licensed and non-licensed work

Regulation 8 of CAR 2012 states that "an employer shall not undertake any work with asbestos unless he holds a licence" issued by the HSE.

The definition of 'asbestos' applies to all forms of asbestos. See Regulation 2(1) for a full list.

'Work with asbestos' is defined in ACOP 27 and includes "its removal, repair or disturbance; any ancillary work and supervising work".

This onerous requirement is only qualified by specific and limited exceptions as set out in Regulation 3(2).

Most asbestos work must be undertaken by a licenced contractor, but any decision on whether particular work is licensable is based on the risk.

Work with asbestos falling into any of the following categories is **licensable work**:

- where worker exposure to asbestos is not "sporadic and of low intensity" (the concentration of asbestos in the air should not exceed $0.6f/cm^3$ measured over 10 minutes)
- where the risk assessment cannot clearly demonstrate that the control limit will not be exceeded, i.e. 0.1 asbestos fibres per cubic centimetre of air ($0.1\ f/cm^3$)
- on asbestos coating
- on asbestos insulation or asbestos insulating board where the risk assessment demonstrates that the work is not short duration work, e.g. when work with these materials will take no more than 2 hours in any 7-day period, and no one person works for more than 1 hour in that 2-hour period.

To be exempt from needing a licence the work must be sporadic and low intensity and be carried out so that the exposure of workers to asbestos will not exceed the legal control limit. It must also meet at least one of the following additional conditions:

- It is a short non-continuous maintenance task, with only non-friable materials
- It is a removal task, where the asbestos containing materials (ACMs) are in reasonable condition and are not being deliberately broken up, and the asbestos fibres are firmly contained within a matrix, e.g. the asbestos is coated, covered, or contained within another material, such as cement, paint, or plastic
- It is a task where the ACMs are in good condition and are being sealed or encapsulated to ensure they are not easily damaged in the future
- It is an air monitoring and control task to check fibre concentrations in the air
- It is the collection and analysis of asbestos samples to confirm the presence of asbestos in a material

CAR 2012 has introduced a third category: **notifiable non-licensed work** (NNLW). This includes works which do not require the employment of a licensed contractor but which nevertheless are considered significant enough by the HSE to require their notification to the HSE and provision of health surveillance and keeping of exposure and health records.

It is the responsibility of the person in charge of the job to assess the ACM to be worked on and decide if the work is NNLW or non-licensed work. This is subjective and will be a matter of judgement in each case, taking into account the nature of the work, the type of asbestos (its friability) and its condition.

The HSE has published guidance giving examples of circumstances which are NNLW works and also which normally will not be NNLW and are thus non licensed work. Refer to the HSE website (www.hse.gov.uk), in particular:

- "Asbestos essentials" Sheet A0 – contains a decision flow chart
- Licensing/notifiable- non-licenced work – gives details of the applicable work categories
- Licensing/asbestos-work categories pdf – chart illustrating different work categories

Notification is made by the employer of the workers. There is no minimum notice period, but notice must be made before the work starts and can only be made on line. The requirements in respect of medical surveillance for workers undertaking notifiable non-licensed work are less stringent than those for licensed work.

Training

ACOP L143 *Work with materials containing asbestos* sets out the detailed requirements.

In addition to the general training requirements for any employee, set out in the Management at Work Regulations, Regulation 10 of CAR 2006 requires every employer to ensure that adequate information, instruction, and training are given to employees who:

- are, or are liable to be, exposed to asbestos, and to their supervisors; and
- carry out work in connection with the employer's duties under these Regulations so that they can carry out that work effectively.

The ACOP, which accompanies CAR 2006, provides a full and detailed list of these training requirements.

The following table lists the three main types of information, instruction, and training.

Table 4.5

Type	For
Awareness training	Those who are liable to be exposed to asbestos while carrying out their normal everyday work, e.g. maintenance staff; electricians; demolition and construction workers; installers of computers, fire, or burglar alarms; and 'construction professionals'.
For non-licensable asbestos work	For example, roofer removing whole asbestos cement sheet in good condition.
For licensable asbestos work	For example, a contractor removing asbestos lagging or asbestos insulating board.

The topics which the training should cover are listed above and should be given in appropriate detail by both written and oral presentation and by demonstration (as necessary). In particular, training is to be in a manner appropriate to the nature and degree of exposure identified by the employer's risk assessment, giving the significant findings of the assessment and the results of any air monitoring carried out, together with an explanation of the findings.

A competent person should give the training and the procedures for providing the information, instruction and training should be clearly defined and documented, and reviewed regularly, particularly when work methods change.

Records should be kept of the training undertaken by each individual. For licensable work, copies should be given to each individual. Refresher training should be given at least every year and more frequently if the work methods, equipment used or type of work changes.

Where non-employees are on the employer's premises, they should also be given adequate information, instruction and training as far as is reasonably practicable.

Management of asbestos

Regulation 4 requires the dutyholder(s) to 'manage' asbestos in 'non-domestic premises' and also, for 'every person' to cooperate with the dutyholder so far as is necessary to enable him or her to comply with his or her duties.

Accredited personnel

Regulations 20 and 21 require, respectively, that only persons accredited as complying with ISO 17025 must be employed to measure the concentration of airborne asbestos fibres and analyse a bulk sample of material to determine whether it contains asbestos.

In addition, anyone issuing a clearance certificate for reoccupation following asbestos removal work is required to meet the relevant accreditation requirements of ISO 17025 and ISO 17020 extended to include all **four stages** of the clearance and not just the air testing part.

Distinguishing between asbestos insulating board and asbestos cement

Because of the relatively higher potential risks to health arising, working with asbestos insulating board (AIB) is subject to different and more stringent controls than working with asbestos cement.

As the two materials are very similar in physical appearance, depending upon the circumstances, it is necessary to distinguish between them. The distinction is one of density, the general rule being that the greater the proportion of cementitious matrix to asbestos, the greater the density. Conversely, the more friable the material, the greater the asbestos content.

ACOP 147 defines 'asbestos cement' as "a material which is predominantly a mixture of cement and asbestos and which in a dry state absorbs less than 30% water by weight". The guidance in the document includes a comprehensive description of the testing procedure.

'Asbestos insulating board (AIB)' is defined as "a flat sheet, tile or building board consisting of a mixture of asbestos and other material except asbestos cement or any article of bitumen, plastic, resin or rubber which contains asbestos, and the thermal or acoustic properties of which are incidental to its main purpose".

It is relatively easy to identify asbestos cement when it is used in preformed components such as corrugated sheeting, tanks or toilet cisterns, but when it is used in flat board form the only sure way of distinguishing the material from AIB is by laboratory analysis. Where there is doubt, the Regulations require that caution be taken, and the material is presumed to be AIB until proven otherwise.

Table 4.6 Summary of CAR 2012

Regulation(s)	Title (in italics) plus explanatory notes
1	*Citation and commencement*
2	*Interpretation*
3	*Application of the Regulations*: see 'Regulation 3(2) exceptions'
4	*Duty to manage asbestos in non-domestic premises*: see separate detailed explanation following this table
5	Identification of the presence of asbestos
6	*Assessment of work which exposes employees to asbestos*: prior to the works, assess the likely level of risk, determine the nature and degree of exposure, and set out steps to control it

Regulation(s)	Title (in italics) plus explanatory notes
7	*Plans of work*: produce a suitable written plan of the proposed works
8*	*Licensing of work with asbestos*: All work except 'Regulation 3(2) exceptions' must only be undertaken by licence holders
9*	*Notification of work with asbestos*: notify the enforcing authority of licensable work (min. 14 days notice)
10	Information, instruction, and training
11	*Prevention or reduction of exposure to asbestos*: prevent exposure to employees as far as is reasonably practicable, and where not reasonably practicable, reduce to lowest level reasonably practicable, both the exposure (without relying on use of respirators) and the number of employees exposed
12	*Use of control measures*: ensure that any control measures are properly used or applied
13	*Maintenance of control measures*: maintain control measures and equipment (keeping records of the latter)
14	*Provision and cleaning of protective clothing*: provide suitable personal protective equipment (PPE) and ensure it is properly used and maintained
15*	Arrangements to deal with accidents, incidents, and emergencies
16	*Duty to prevent or reduce the spread of asbestos*: where not reasonably practicable, reduce to lowest level reasonably practicable
17	*Cleanliness of premises and plant*: keep asbestos working areas and plant clean and thoroughly clean on completion; four-stage clearance procedure and certification for reoccupation
18*	*Designated areas for asbestos work*: 'asbestos areas' where any employee would be liable to be exposed to asbestos and 'respirator zones' where control limit liable to be exceeded
19	*Air monitoring*: monitor exposure of employees to asbestos

Regulation(s)	Title (in italics) plus explanatory notes
20	*Standards for air testing and site clearance certification*: see 'Accredited personnel'
21	*Standards for analysis*: of bulk samples, see 'Accredited personnel'
22*	*Health records and medical surveillance*: of employees liable to be exposed to asbestos
23	Washing and changing facilities
24	*Storage, distribution and labelling of raw asbestos and asbestos waste*: if received, dispatched from, transported or distributed, waste must be in sealed, labelled, appropriate bags or containers
25–28	Prohibitions and related provisions
29–34	*Miscellaneous*: exemptions, etc. and defence 'must have taken all reasonable precautions and exercised all due diligence'

Regulations marked with an asterisk do **not** apply in whole or part where the work falls within any of the circumstances set out in 'Regulation 3(2) exceptions'.

Duty to manage asbestos in non-domestic premises (Regulation 4)

The dutyholder responsible for the management of asbestos in non-domestic premises as set out in Regulation 4(1) is every person who has, by virtue of a contract or tenancy, an obligation for its repair or maintenance, or, in the absence of such, control of those premises or of access to or egress from the premises.

This includes those persons with any responsibility for the maintenance, or control, of the whole or part of the premises.

When there is more than one dutyholder, the relative contribution required from each party in order to comply with the statutory duty will be shared according to the nature and extent of the repair obligation owed by each.

This regulation does not apply to 'domestic premises', namely a private dwelling in which a person lives, but legal precedents have established that common parts of flats (in housing developments, blocks of flats and some conversions) are not part of a private dwelling.

The common parts are thus classified as 'non-domestic' and therefore Regulation 4 applies to them, but not to the individual flats or houses in which they are provided.

Typical examples of common parts are entrance foyers, corridors, lifts, their enclosures and lobbies, staircases, common toilets, boiler rooms, roof spaces, plant rooms, communal services, risers, ducts, and external outhouses, canopies, gardens, and yards.

The regulation does not however apply to kitchens, bathrooms, or other rooms within a private residence that are shared by more than one household, or communal rooms within sheltered accommodation.

The duties are identified in the following table.

Table 4.7

Subject	Requirement
Cooperate (see 'Duty to cooperate')	Cooperate with other dutyholders so far as is necessary to enable them to comply with their Regulation 4 duties.
Find and assess condition of asbestos containing materials (ACMs)	Ensure that a suitable and sufficient assessment is made as to whether asbestos is or is liable to be present in the premises and its condition, taking full account of building plans or other relevant information, the age of the building and inspecting those parts of the premises which are reasonably accessible. (Must presume that materials contain asbestos unless strong evidence to the contrary.) (See HSG264 *Asbestos: The survey guide* for guidance on asbestos surveys.)
Review	Review assessment if significant change to premises or suspect that it is no longer valid and record conclusions of each review.

Subject	Requirement
Records	Keep an up to date written record of the location, type (where known), form and condition of ACMs.
Risk assessment	Where asbestos is or is liable to be present assess the risk of exposure from known and presumed ACMs.
Management plan (see also below)	Prepare and implement a written plan, identifying those parts of the premises concerned, specifying measures for managing the risk, including adequate measures for properly maintaining asbestos or, where necessary, its safe removal.
Provide information to others	Ensure the plan includes adequate measures to ensure that information about the location and condition of any asbestos is provided to every person likely to disturb it and is made available to the emergency services.
Review and monitor	Regularly review and monitor the plan to ensure it is valid and that the measures specified are implemented and that these are recorded.

Management plan

The management plan is an important legal document that, in addition to its health and safety significance, will be required to be made available to, and inspected by, a variety of interested parties. The absence of such a document may therefore have significant financial implications or affect the liquidity of the premises as an asset.

A plan is not required when the assessment whether asbestos is present or is liable to be present in the premises confirms that it is not. For example, the building is post-2000, and there is confirmation from the project team that asbestos has not been used in its construction.

Nevertheless, a record must be kept of the assessment carried out and its conclusion to show to an inspector or prospective purchaser or occupant. The dutyholder owns and is responsible for the safekeeping of the plan, however, he or she is obliged to make the information available 'at a justifiable and reasonable cost' to anyone who is likely to disturb asbestos, and this includes new owners or occupants. (See HSE publication *A comprehensive guide to managing asbestos in premises* HSG227.)

Duty to cooperate

Every person has a duty to cooperate with the dutyholder so far as is necessary to enable the dutyholder to comply with his or her duties under Regulation 4. This includes the landlord, tenants, occupants, managing agent, contractors, designers, and planning supervisor.

The possible scenarios envisaged by the ACOP include:

- anyone with relevant information on the presence (or absence) of asbestos; and
- anyone who controls parts of the premises to which access will be necessary to facilitate the survey and management of asbestos (i.e. its removal, treatment or periodic inspection).

Cooperation does not extend to paying the whole or even part of the costs associated with the management of the risks of asbestos by the dutyholder(s), who must meet these personally.

Where there is more than one dutyholder for premises, the costs of compliance will be apportioned according to the terms of any lease or contract determining the obligation to any extent for repair and maintenance of the premises.

If there is no such documentation then the apportionment of costs will be based on the extent to which parties exercise physical control over the premises.

In the final analysis, the courts will decide financial responsibility using the principles outlined above. Guidance in the

ACOP states that architects, surveyors, or building contractors who were involved in the construction or maintenance of the building and who may have information that is relevant 'would be expected to make this available at a justifiable and reasonable cost'.

The duty to cooperate is not subject to any limitation or exclusion, thus there is an obligation to do whatever is necessary to cooperate with the dutyholder.

Short lease tenants, licensees or other occupants who control access but do not have any contractual maintenance liabilities would be required to permit the landlord access to fulfil his or her duties.

In May 2009, RICS published the second edition of its guidance note *Asbestos and its implications for members and their clients* (ISBN 978 1 84219 450 8). At the time of going to print it is anticipated that a third edition will be issued in 2011 to take account of HSG264.

HSG264 Asbestos: The survey guide (guidance on asbestos surveys)

Issued by HSE in January 2010, this replaces and expands on MDHS 100 *Methods for the determination of hazardous substances – Surveying, sampling, and assessment of asbestos-containing materials*, issued July 2001. The key changes are as follows:

> **Categories of survey**: previously there were three types of asbestos survey – this has been reduced to two, 'Management survey' and 'Refurbishment and demolition survey'. The former is that necessary to meet the CAR Regulation 4 duties during normal occupancy, including foreseeable maintenance and installation. The latter is required **before** any refurbishment or demolition and for all work which disturbs the building fabric in areas where the management survey has not been intrusive.

Competence and quality assurance procedures: clarification on the technical competence of asbestos surveyors 'strongly recommending the use of accredited or certified asbestos surveyors', requiring the use of 'adequate quality management systems' and that the survey is carried out in accordance with the survey guide.

Survey planning: emphasis on the importance of pre-inspection planning with a survey strategy for both non-domestic and domestic properties, with the aim of reducing caveats and ensuring the completeness of the survey.

Polychlorinated biphenyls and buildings

Trevor Rushton

Polychlorinated biphenyls (PCBs) are mixtures of up to 209 individual chlorinated organic compounds. Their commercial use began in 1929 and they have since been used as a component in insulation, as cooling fluids in transformers, as dielectric fluid in capacitors, in hydraulic fluids, in grouting and sealants, and as plasticisers in paints. Since their introduction, over 1m tonnes of PCBs have been produced worldwide (OSPAR Commission, 2001). PCBs are now banned, although in certain cases, equipment containing PCBs can still be used. Equipment that contains more than 5 litres of fluid containing PCBs (i.e. more than 5 litres of fluid that has a PCB concentration of more than 0.005%) is classed as 'contaminated equipment'.

Health implications

Concerns have been raised since the 1970s regarding PCBs' toxicity, persistence, and tendency to bio-accumulate – once they are in the environment or have been ingested by humans, it is very difficult to dispose of them. They can adversely affect reproduction, immune and nerve systems and children's ability to learn. Extensive exposure has been proven to result in skin problems, lung and liver damage, and disruption to hormone and enzyme systems.

Poisoning from PCBs affects an animal's ability to reproduce. These effects have been observed among seals, minks, guillemots and sea eagles. When the PCB content in the environment was at peak levels, these species were adversely affected and population levels dropped. Studies have also shown that pregnant women who ate fish from lakes (implying a moderate exposure to PCBs) have given birth to children with reduced learning abilities.

PCB audits and what to look for

PCBs were widely used as a dielectric fluid in electrical transformers and capacitors up to 1977 when manufacture of these materials ceased. Buildings constructed prior to this may therefore harbour materials or components containing PCBs.

Movement joints containing PCBs can be found, for example, in facades (dilatation joints), around facade elements, around doors and windows, hidden underneath thresholds, in connections between facades and balconies, and also hidden from sight, for example, behind steel sheet claddings. PCBs in plastic-based floor coverings, especially with non-slip qualities, were used in industrial premises and institutional (large-scale) kitchens. PCBs can also be found in sealed insulating glass units (consisting of two glass panes with a separating profile sealed with a jointing compound) and in capacitors used in electric fittings.

When undertaking a PCB survey, the surveyor should use adequate personal safety equipment (such as disposable gloves) and be aware that PCBs are highly contaminating. All tools used for removing materials must be carefully cleaned with acetone after each sample has been collected and the

gloves disposed of. The samples must be completely sepa-
rated from each other using an aluminium foil wrapping
and each sample should then be placed in a plastic bag and
properly marked before being sent for analysis. Before undertaking the survey, it is useful to make a plan of the building
and mark where the samples will be taken.

Initial PCB decontaminations were not always successful.
For example, new jointing material was found to have been
contaminated by PCBs from materials surrounding the new
jointing compound. In Scandinavian countries this led to the
development of extensive guidelines on decontaminating
joints. The guidelines describe, for example, how surrounding
materials, such as concrete facade elements, have to be
partly ground down to remove the PCB.

Registration and disposal

Whilst it is not illegal to continue to operate equipment
containing some PCBs (contaminated equipment or CE),
the operator/owner must register any with the Environment
Agency, including CE that:

- has a legal use (e.g. transformers with a PCB concentration below 0.05%), and
- doesn't have a legal use, but has not been disposed of yet.

An annual fee is payable upon registration and clear warning
notices must be displayed.

Contaminated equipment must be disposed of in a way that
destroys the PCBs. If this is not possible, the owner may apply
for permission to store it underground.

Further information
- www.gov.uk/guidance/polychlorinated-biphenyls-pcbs-
registration-disposal-labelling   http://www.netregs.
org.uk/library_of_topics/materials__equipment/

more_hazardous_materials_topic/polychlorinated_biphe-
nyls.aspx
- Ospar Commission: www.ospar.org
- *Investigating Hazardous and Deleterious Building
  Materials*, Rushton, T., RICS Books, 2006.

High alumina cement concrete (calcium aluminate cement)

Trevor Rushton

In 1925, cement producer Lafarge commenced the UK manufacture of high alumina cement (HAC) to provide concrete
that would resist chemical attack, particularly for marine
applications. This cement developed high early strength,
although its relatively high cost prevented extensive use.

During the late 1950s and 1960s the main use of HAC was
for the manufacture of precast prestressed components
which could be manufactured quickly, therefore offset-
ting the additional cost of the material. However, HAC loses
strength with age and can become vulnerable to chemical
attack.

The earliest UK failures were experienced during 1973–74
when several school roofs collapsed. In 1976 HAC concrete
was banned for structural use, although new uses are now
becoming established under the name calcium aluminate
cement.

Problems with HAC

HAC concrete undergoes a mineralogical change known
as 'conversion'. During this process the concrete increases
in porosity, which in turn results in a loss of strength and a
reduction in resistance to chemical attack. The higher the
temperature during the casting of the concrete, the more
quickly conversion takes place.

The relationship between conversion and strength is complex, however the strength of highly converted concrete is extremely variable and is substantially less than its initial strength. Typically the original design strength of 60N/mm^2 may be reduced to 21N/mm^2.

Highly converted HAC concrete is vulnerable to acid, alkaline and sulphate attack. For this to take place, water as well as the chemicals must have been present persistently over a long period of time at normal temperatures. Chemical attack is usually very localised in nature and the concrete typically degenerates to a chocolate brown colour and becomes very friable. In a warm and moist environment there is the possibility of a serious reaction known as 'alkaline hydrolosis' occurring where high alkali levels may be present as a consequence of the use of certain types of aggregate or where alkalis may have ingressed from plasters, screeds and woodwool slabs. Alkaline hydrolysis is characterised by white powdery deposits and a severe loss of strength and integrity.

Given the sensitivity to moisture the greatest risk thus lies in the use of HAC concrete in roof members. It is therefore important to appraise the condition of the concrete and any waterproof coverings before making any formal judgment as to the remedial work required.

Investigation

There are three generic stages in an investigation, namely:

Stage 1: identification
Stage 2: strength assessment
Stage 3: durability assessment

A document known as the BRAC (Building Regulations Advisory Committee) rules is often used in the assessment of HAC components of certain standard types. The rules were first produced in 1975 and were reviewed recently; they are now published by the Building Research Establishment (BRE), albeit the guidance is essentially the same.

Assuming that chemical tests have identified HAC, a Stage 2 strength assessment is required to determine if the precast concrete members have sufficient structural capacity, even at the reduced fully converted strength, to safely withstand the applied loading. The BRAC rules recommend that the assessment be based upon an assumed concrete strength of $21N/mm^2$.

In other words there is no need to work out the degree of conversion that has taken place. The strength assessment requires the section properties of the beam to be established. Thereafter the structural strength of the element can be calculated. In cases where the section properties are unknown, or cannot be determined by investigation, then assessments are limited to determining the concrete strength using near-to-surface tests.

The analysis will only work if the concrete is unaffected by alkaline hydrolysis, since the strength of the concrete could be below the $21N/mm^2$ assumed value.

Note that the calculation methods outlined in the BRAC guidance will only give accurate results for specific types of construction. Floors constructed using composite methods may seemingly fail to satisfy the calculations and yet perform perfectly well. In these circumstances in-situ load testing may give a more reliable result.

A Stage 3 durability assessment is required to determine the long-term durability and risk of chemical attack and reinforcement corrosion. Testing can be undertaken to determine the presence of alkalis and sulphates. Laboratory testing may be supplemented by a detailed visual inspection and the removal of lump samples for petrographic examination. A durability assessment should also include a visual examination of the reinforcing steel where lump samples are removed. In recent years it has been found that HAC is less durable than members containing ordinary Portland cement.

Loss of strength is only one of the issues that affects HAC concrete, and to conduct a full appraisal of an HAC structure it is necessary to consider three issues:

- Loss of strength associated with the process of conversion
- Carbonation
- Damage by chemical attack, mainly sulphates or alkalis

Assessment

Putting all of this in context, however, there have been no recorded instances in the UK of a failure of a floor incorporating HAC concrete. It is important to know that in the case of the original historic failures, manufacturing faults were eventually discovered. The greatest reduction in strength occurred where a high water content was present during the period of mixing and there were high temperatures during curing.

Of the five failures or new failures of roof constructions, two did not directly involve the quality of the concrete, one was aggravated by chemical attack and two were apparently due to defective concrete which should have been rejected at the time of casting. On no occasion has weakening of the concrete due to conversion been the sole cause of failure.

Recent moves to rehabilitate HAC have involved subtle changes in the Approved Documents to the Building Regulations, while BS EN 14647: 2005 contains an appendix listing the principles for the correct use of calcium aluminate cements (formerly known as HAC).

Further information
Investigating hazardous and deleterious building materials, Rushton T RICS Books, 2006.

BRE Report 451 (The BRAC Rules), Building Research Establishment, 2002.
BRE Special Digest 3: *HAC concrete in the UK: Assessment, durability management, maintenance and refurbishment,* Building Research Establishment, 2002.

Calcium chloride

Trevor Rushton

Calcium chloride may be present in reinforced concrete as a result of its inclusion as an accelerator, by contamination from de-icing salts, or from the use of unwashed or poorly washed marine aggregate. Sufficient levels of chloride may result in 'chloride-induced corrosion' which can be more difficult to deal with than corrosion caused purely by carbonation. When chloride corrosion does occur, its effects may be wide ranging, including a reduction in structural capacity.

Natural alkalinity of concrete

Steel does not corrode when embedded in highly alkaline concrete, despite high moisture levels in the concrete, because a passive film forms on the steel and remains intact as long as the concrete surrounding the bar remains highly alkaline (pH above 12.6).

Chloride-induced corrosion

The use of calcium chloride as an accelerating additive at the time of mixing was popular during the 1950s and 1960s. It was used in precasting yards to speed up the reuse of expensive moulds and was used on site during cold weather to increase the rate of gain in strength. The use of calcium chloride additive was banned in 1977. Corrosion may occur in concrete that contains sufficient chlorides even if it is not carbonated or showing visible signs of deterioration.

Chloride ions exist in two forms in concrete, namely free chloride ions, mainly found in the capillary pore water, and combined chloride ions which result from the reaction between chloride and the cement hydration process. These occur in proportions that depend on when the chloride entered the concrete. If chloride was introduced at mixing, for example, as an accelerator, approximately 90% may form harmless complexes leaving only 10% as free chloride ions. If, on the other hand, seawater or de-icing salts penetrate the surface of the concrete, the ratio of free to combined chloride may be 50:50.

All aggregates used commonly for concrete mixing contain a background level of chlorides, usually less than 0.06% by weight of cement. The presence of calcium chloride cast-in within the mix usually attracts a chloride level significantly greater than 0.4% by weight of cement. Ingressed chlorides through the outer surface of the concrete are variable in nature.

However, the use of de-icing salts on, for example, external staircases and balconies is popular and may result in localised high concentrations of chlorides in excess of 1.0% by weight of cement. Concentrations of ingressed chlorides on the top surface of a car park deck may typically occur up to 3 or 4%.

The presence of free chloride ions within the pore structure of the concrete interferes with the passive protective film formed naturally on reinforcing steel.

The corrosive effect of chlorides is significantly affected therefore by the presence of free chlorides. The effects of chlorides are classified in terms of risk of corrosion because in certain conditions even low levels of chloride may pose some risk. The permissible level of chloride added at mixing specified in BS 8110 is 0.4% by weight of cement. For prestressed concrete the level is lower at 0.06%. The overall effect of reinforcement corrosion caused by chlorides must therefore be considered with the depth of reinforcement and the depth of carbonation.

Chloride-induced corrosion results in localised breakdown of the passive film rather than the widespread deterioration that occurs with carbonation. The result is rapid corrosion of the metal at the anode, leading to the formation of a 'pit' in the bar surface and significant loss of cross sectional area. This is known as 'pitting corrosion'. Occasionally a bar may be completely eaten through.

Chloride-induced corrosion may occur even in apparently benign conditions where the concrete quality appears to be satisfactory. Even if there is poor oxygen supply, reinforcement corrosion may still take place. Failure of reinforcement may therefore occur without any visual sign of cracking or spalling that would otherwise occur following the formation of expansive rust.

The presence of calcium chloride is further exaggerated by the presence of deep carbonation. Carbonation releases combined chlorides into solution to form free chloride ions, thus increasing the likelihood of corrosion. For this reason, many properties built approximately 20–30 years ago may only now start causing problems.

Risk assessment

Guidance on assessing the risk of reinforcement corrosion is provided by the Building Research Establishment (BRE) in Digest 444 Part 2. Here the risk of corrosion for structures of various ages is presented in a range from negligible to extremely high risk. Factors affecting the risk assessment are either a dry or damp environment, the depth of carbonation and of course the level of chlorides present. The digest includes risk assessment tables indicating the risk of corrosion activity according to the age of the building, whether in a wet or dry environment and whether the concrete is carbonated or uncarbonated.

Repair

The successful repair of chloride-induced corrosion is notoriously difficult because of the tendency for new corrosion cells to form at the boundary of the repair. This mechanism is called 'incipient anode effect' and should be minimised by removing, wherever possible, all concrete with significant chloride contamination. In recent years the introduction of proprietary sacrificial zinc anodes embedded within the patch repair and attached to the reinforcement can help to reduce this effect. For high levels of chlorides and long-term protection this may not be sufficient.

For heavily chloride-contaminated structures, particularly car parks, the only tried and tested long-term solution is cathodic protection. Cathodic protection is a means of corrosion protection whereby the potential of a metal structure is made more negative in order to decrease corrosion rates.

The cost and complexity of installing cathodic protection is not usually warranted within building structures. A variation of cathodic protection is desalination – a short-term process using higher current densities than cathodic protection. Desalination gained some popularity in the 1990s but its use has since declined.

An alternative treatment is the application of surface-applied coatings that then migrate into the concrete. Migrating corrosion inhibitors have found some success, but there is some nervousness about their effectiveness in view of the difficulty of checking site work; the inhibitors must be applied under strict conditions.

Repair of concrete is now covered by BS EN 1504. See also *Concrete Society Technical Report 69*.

Further information
   Corrosion of steel in concrete, BRE Digest 444 Part 2, BRE, 1997.

Investigating hazardous and deleterious building materials, Rushton, T., RICS Books, 2006.
An introduction to electrochemical rehabilitation techniques, Drewitt, J., Broomfield, J., Corrosion Prevention Association, Technical Note 2, 2011.

Alkali aggregate reactions

Trevor Rushton

Alkali aggregate reactions (AAR), of which alkali silica reactions (ASR) are only one variant, are relatively uncommon in building construction in the UK and tend to occur in civil engineering structures rather than conventional buildings. Another, even rarer chemical reaction, is alkali-carbonate reaction (ACR).

Alkali silica reactions – the problem

Concrete is a highly alkaline material: it follows that pore water contained within the concrete will also be highly alkaline. ASR occurs when, given the correct combination of conditions, the pore water can react with certain types of aggregate to produce a gel.

The gel absorbs water, expands and can cause the concrete to crack or disrupt. Sometimes a pattern of 'map' cracking occurs but in others small 'pop-outs' can form – rather like concrete with acne. The durability of the concrete can thus be compromised and in extreme cases the tensile strength of the concrete component can be reduced.

ASR in concrete can be very damaging, sometimes resulting in structural failure and the need to demolish a building or to undertake significant structural repair works. However, the problem is infrequent and since the introduction guidance in BS 8500, the British Standard for Concrete — Complementary

British Standard to BS EN 206-1 it is believed that the frequency of attack in new buildings has been diminishing since the early 1990s.

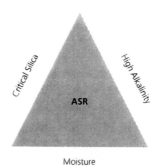

Moisture

Figure 4.1 ASR Triangle.

For ASR to occur, three factors must be present. Remove one and ASR will not take place:

- Critical silica in the aggregate – the level of silica will affect reactivity, aggregates such as chert can be problematic
- Sufficient moisture – water being a critical component
- High alkalinity, either from the cement or from other external sources

Most of the aggregates used in the UK are considered to be of 'normal' reactivity rather than high reactivity, although there are some troublesome types in the southwest.

The Building Research Establishment (BRE) and the Concrete Society have prepared guidelines and flow charts to assist the designer in the selection of appropriate measures to reduce the risk of ASR.

In order to check whether the alkali content of the cement is above the critical level, chemical tests are needed. When dealing with the total alkali content of concrete it is usual to consider the equivalent sodium oxide in the concrete expressed as kilogrammes per cubic metre (kg/m^3). By adopting this 'sodium oxide equivalent' scale, cement can be classified according to low, moderate or high alkalinity. High alkalinity cement means a 0.75% sodium oxide equivalent or greater.

A combination of normal reactivity aggregate with high alkalinity cement is permissible, if the total alkalinity of the concrete is kept within certain limits. However, good construction practice dictates that for normally reactive aggregates, the total alkali content of the concrete should not exceed 3.0kg/m^3.

What if ASR occurs?

A popular technique for the identification of ASR is the examination of thin sections of concrete, using a petrographic microscope. Polished sections of concrete can, in the alternative, be examined by scanning electron microscopy (SEM).

Concrete Society Report 30 and Part 4 of BRE Digest 330 are mainly concerned with the selection of materials to reduce the risk of ASR occurring. The available records seem to point to a low risk of the development of ASR, but remember that the third component necessary for the formation of a reaction is sufficient moisture.

If the concrete was reinforced, then any potentially expansive reactions in the concrete could be restrained (the degree of restraint depending upon the arrangement of the reinforcement).

A risk assessment based approach to analysis is advocated – this should take into account not only whether the concrete is wet or dry, or is restrained with reinforcement or not, but also the consequences of failure – the structural significance of a component. The structural significance of an element is a function of the consequences of its failure. These are judged to be slight or significant as defined below:

- **slight**: the consequences of structural failure are either not serious or are localised to the extent that a serious situation is not anticipated
- **significant**: if there is risk to life and limb or a considerable risk of serious damage to property

Further information
BRE Digest 330, *Alkali-silica reaction in concrete.*
Structural effects of alkali-silica reaction: Technical guidance on the appraisal of existing structures, Institute of Structural Engineers, 1992.
Concrete Society Report 30, *Alkali-silica reaction – Minimizing the risk of damage* (3rd edition), TR30, 1999.
Investigating hazardous and deleterious building materials, Rushton, T., RICS Books, 2006.
Alkali silica reaction, Fact Sheet 4, British Cement Association, 2006.
Concrete – Complementary British Standard to BS EN 206. Specification for constituent materials and concrete (2015).

Concrete repair
Trevor Rushton

Concrete is generally a durable material, but poorly designed or constructed concrete structures can require extensive repair. In particular, many concrete buildings constructed in the 1960s and 1970s are now in poor condition. Deterioration can occur to both the concrete and the embedded reinforcement.

Deterioration of reinforcement due to carbonation can often be identified by visible rust staining or spalling of concrete. However, concrete containing excessive chlorides can lead to severe corrosion of reinforcement without any corresponding visual indication of the problem on the surface. Chloride-induced corrosion can occur where the original concrete mix included chloride-containing additives. However, it is also particularly prevalent in car park structures where de-icing salts from vehicles are present. The correct choice of repair technique for each type of corrosion is important; there is the possibility of inappropriate repair actually accelerating the corrosion process.

Concrete repair work is covered by the ten-part British Standard EN 1504 *Products and systems for the protection and repair of concrete structures – Definitions requirements, quality control and evaluation of conformity*.

The Standard is comprehensive and includes guidance on assessment of condition, identification of causes of deterioration, options for protection and repair, and specification of maintenance requirements following protection and repair.

The Standard introduces in Table 1 of Part 9, the concept of 'Principles and methods for protection and repair of concrete structures'. There are 11 principles and these are set out in detail in *Annex A: Guidance and background information*.

Before undertaking extensive repair of concrete structures, consideration should be given to the likely rate of ongoing deterioration and the required life of the structure, so that the cost effectiveness of different protection and repair strategies can be considered.

The possibility of adverse effects from repair methods and the consequences of interaction between them must be considered before specifying repairs.

Many concrete repair techniques rely on the use of proprietary products or systems, and the work is often done by specialist contractors. The product manufacturers in particular can provide valuable advice in the correct use of their systems, but it is important to obtain independent advice so that the merits of different repair techniques can be compared.

Parts 9 and 10 of the Standard are the most relevant to specification and execution of repair work on site. Part 9: *General principles for use of products and systems* sets out common causes of deterioration which should be considered before investigating or treating deteriorated concrete.

Section A.6.2 of the Standard also draws attention to the possibility of adverse effects of chosen repair methods and the consequences of interaction between them.

Note: BS EN 1504 does not deal with all concrete repair methods, for example, electrochemical chloride extraction and electrochemical re-alkalisation of carbonated concrete are not covered.

Successful repairs

Successful repairs are more likely to be achieved by following a careful plan of action:

- Assess the condition of the structure – what are the main agencies of damage – chloride, carbonation, lack of adequate cover, etc.?
- Consider the options for action – repair, demolition, make safe, do nothing?
- Decide upon the methodology – restore protection, barrier coating, restore integrity, patch repair, overcoating?
- Select materials based upon required performance standards.
- Undertake the repair.
- Establish and define maintenance and control strategy for the future.

Further information
BS EN 1504 Parts 1-10 – British Standards Institution
The route to a successful concrete repair. Guidance Note 1 (2nd edition), Concrete Repair Association, 2009.

Corrosion of metals
Trevor Rushton

Metals used in building are almost invariably the product of large amounts of energy used to transform raw materials (ores) into a finished product. The transformation from ore to metal creates an inherently metastable material which, under the influence of water and oxygen, will gradually revert (corrode) to its original form, releasing energy as it does so.

'Corrosion' is an electrochemical process in which a metal reacts with the environment in which it is located to form an oxide or other compound. Energy is released in the form of corrosion currents – these vary according to the metal.

Aside from gold, metals exposed to air develop an oxide film, a layer which is 30–100 Angstroms thick (about 15–50 atoms thick). The passive layer serves to protect the metal from corrosion; it greatly reduces the rate of passage of metal ions from the surface. In this condition the metal is described as 'being in a passive state'.

If the metal is then exposed to an aqueous solution (electrolyte), the oxide film will tend to dissolve. Once this occurs, and the metal is exposed, it is termed active. In its active state, positively charged ions tend to pass from the metal into the electrolyte. (An 'ion' is an atom or group of atoms that carries a positive or negative charge as a result of having lost or gained one or more electrons. Ions with a positive charge are called 'cations'; ions with a negative charge are called 'anions'.)

Metallic ions, because they are formed from atoms that have lost electrons, are positively charged – when an atom or ion loses electrons it is said to have been 'oxidised'.

The location of the oxidisation process is described as the anode. Here, the oxidisation of the metal is termed 'an anodic reaction'. This flow creates an increase in the potential difference between the metal and the solution – the potential of the metal.

While the process of oxidisation continues, the deposition of dissolved metal ions from the solution occurs at the cathode. The corrosion current between the anode site and the cathode site consists of electrons flowing within the metal and ions flowing within the electrolyte. The electrolyte does not have to be liquid water – for example, the corrosion cells that form within reinforced concrete rely upon the small amount of moisture that remains in the pores of the concrete.

Corrosion is usually initiated by one or a combination of:

- a difference in the electrical potential of two metals in contact with an electrolyte – a galvanic couple;
- variations in the metallurgical state of the metal at different points on its surface; and/or
- local environmental differences, for example, variations in oxygen supply.

The following very broad categories of corrosion exist.

Table 4.8

Uniform (general) corrosion	Accounts for about 30% of all corrosion-related failures. Occurs over the surface of the metal as a whole and is made up of a multitude of randomly positioned, microscopically sized individual corrosion cells. This form of corrosion occurs when the passive layer has been damaged or dissolved, or where it ceases to offer protection. General corrosion can occur quite rapidly and is usually accompanied by the deposition of the products of corrosion on the surface – in the case of ferrous metals the deposition is rust. Rust does not protect the metal against corrosion, but other oxide forms can, for example, the tarnishing of silver or the green patina associated with the corrosion of copper.
Atmospheric corrosion	Cycles of alternate wetting and drying can create conditions favourable to the corrosion of metals. When the atmosphere is contaminated with pollutants such as sulphur dioxide or chlorides, corrosion rates can be greater. Exposure to this form of corrosion is not confined to external situations; if the internal environment is favourable, corrosion can still occur.

Bimetallic or galvanic corrosion	A consequence of two metals in contact with an aqueous solution
Microbiological corrosion	This term covers a variety of corrosion types initiated mainly by the presence of and/or activities of micro-organisms in biofilms that have formed on the surface of the corroding material. The formation of biofilms, sludge or other deposits can be particularly harmful as it can precipitate crevice corrosion, destroy corrosion inhibitors and directly influence a corrosion cell. A number of microbiological organisms have been associated with corrosion damage in water systems. These micro-organisms can influence corrosion by effects such as the creation of differential aeration cells, the production of corrosive mineral and organic acids, or ammonia, and chemical reductions. Four environmental conditions are required for this type of corrosion damage to occur: • metals (the host location); • nutrients; • water; and • oxygen (although some types of bacteria need only very small amounts of oxygen).
Localised corrosion – pitting corrosion, crevice corrosion, filiform corrosion	Likely to cause very serious effects. One form of localised corrosion is called 'pitting corrosion' and can lead to the total or at least significant loss of section of plain and low alloy steel reinforcement without the formation of expansive rust and the tell-tale signs of concrete spalling. The problem often affects those metals that were selected because of their passive nature – for example, stainless steels, nickel alloys or aluminium alloys. Pitting corrosion penetrates perpendicular to the surface and is mainly associated with particularly aggressive ions such as chloride and fluoride, both of which can initiate in the absence of oxygen. Filiform corrosion can be found in the form of long filaments of growth between aluminium and its particular paint coating. The depth of penetration of this corrosion is generally found to be 10–20μm and is therefore considered to be cosmetic rather than structural. The deterioration process is primarily caused by a break or defect in the coating itself and is most commonly found in moist or wet environments (which may be internal or external) where sections of coated aluminium have been cut to form joints, or sections of cladding have been cut, drilled or damaged. Even the most microscopic marks or breaks in the paint effectively cause the formation of a battery circuit between the exposed metal and adjoining sections which are protected from the air. The corrosion takes the appearance of a worm trail as the 'battery cell' moves along the surface of the aluminium underneath the paint.

Environment assisted cracking – hydrogen embrittlement, stress corrosion, liquid metal assisted cracking and fatigue	Environment assisted cracking can produce catastrophic failures in structural metals. The term is used to describe two prime methods of failure: stress corrosion cracking and hydrogen embrittlement. Hydrogen embrittlement has been blamed for several high-profile failures in recent years; this mainly affects high-grade steels whereby hydrogen atoms (both the most common and the smallest) are able to penetrate the surface of the metal and become entrapped by the subsequent plating or finishing process. The hydrogen can then migrate to regions of high stress causing sudden and unplanned failure. The problem can be mitigated by scrupulous attention to detail during manufacture; this may be achievable by mainstream suppliers, but cheap foreign imports could be unreliable.
Flow enhanced corrosion	The flow of a corrosive fluid across the surface of a metal (for example, the inside surface of a pipe) can lead to rapid erosion of the passive film in localised areas. A combination of erosion due to turbulent flow and corrosion can lead to extremely high pitting rates and pipe failure. The soft alloys of copper, aluminium and lead are particularly prone to this form of corrosion.

Bimetallic corrosion

Corrosion will often occur when two different metals are immersed in water containing an electrolyte such as salt or acid. Similar problems can also exist when dissimilar metals may not be in contact but are connected electrically. Metals submersed in seawater are particularly at risk. Detailed guidance on the subject may be found within BS PD 6484:1979, which contains a series of tables from which it is possible to identify the risk of additional corrosion of one metal in contact with another. However, the process of corrosion is complex and depends upon a number of variable factors and not just the presence of dissimilar metals: relative area, temperature and exposure to oxygen will all affect corrosion rates.

Some basic points:

If a metal is immersed in a conducting liquid (the electrolyte), it will take up a conducting potential. Different metals assume a different potential – this is recorded in the following electrolytic scale.

Electrolytic scale

Anodic end (corrosion) −ve, (less noble metal)
Magnesium
Aluminium
Duralumin
Zinc
Cadmium
Chromium iron (active)
Chromium-nickel-iron (active)
Soft solder
Tin
Lead
Nickel
Brasses
Bronze
Copper
Chromium iron (passive)
Chromium-nickel-iron (passive)
Silver solder
Silver
Gold
Platinum
Cathodic end (protected to the detriment of the anodic metal) +ve, (more noble metal)

Figure 4.2

If two different metals are immersed in the electrolyte a current will flow from the more positive metal (the cathode) towards the more negative metal, the anode. This current flow arises because both metals have different electrical potentials. Standard texts often state that the farther apart the metals are on the electrolytic scale, the greater risk of corrosion. The extent, or severity, of the corrosive action is proportional to the distance of separation of the metals in the list. This is an oversimplification of the position when it comes to the selection of metals during the design of buildings and services, as corrosion rates will also depend upon a number of other factors. For example, even metals that are potentially highly reactive can be used successfully together given the correct precautions.

Seawater is highly conductive, and generally this will permit a greater electrical flow. A high flow produces more corrosion to the anode. By contrast, fresh water is generally less conductive. Rainwater, if polluted, can be moderately conductive and so exposure to a wet environment in an industrial area can be more damaging than in a rural area. Exposure to marine conditions can produce highly conductive conditions.

Surface contamination can absorb moisture from the atmosphere even when condensation does not form and so the fact that two metals are not immersed in water does not necessarily prevent corrosion from occurring. Metals buried in soil perform in much the same way as those that are immersed in water.

Corrosion rates can be affected by the area of the anode or cathode relative to the other. For example, a large cathode area and a small anodic area could result in a more severe condition.

Protection against bimetallic corrosion

Actions which may help prevent bimetallic corrosion include:

- insulating the metals from one another;
- applying a metallic coating;
- applying a non-metallic coating;
- using a jointing compound capable of excluding water;
- applying a paint coating; and
- applying sacrificial protection, e.g. cathodic protection.

The data within PD 6484 may be simplified and reproduced as in the following table, which should be treated as a guide only. If a particular combination is critical refer to the more complete information in PD 6484: 1979. The table shows the performance of each metal in contact.

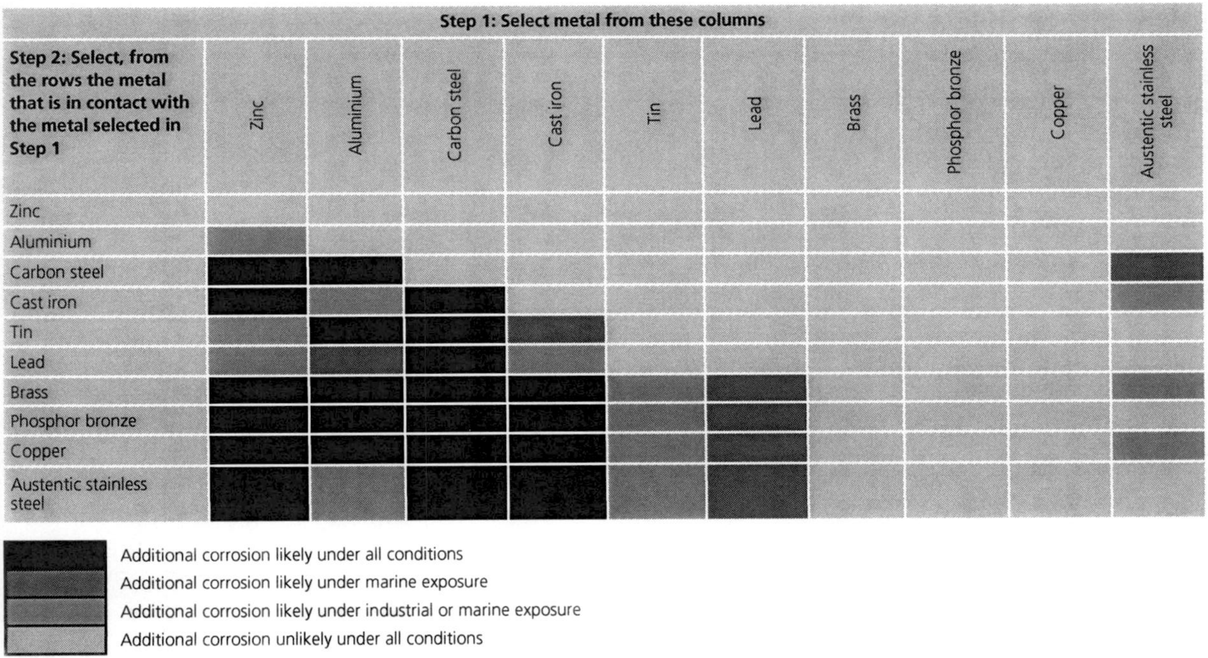

Step 1: Select metal from these columns

Step 2: Select, from the rows the metal that is in contact with the metal selected in Step 1	Zinc	Aluminium	Carbon steel	Cast iron	Tin	Lead	Brass	Phosphor bronze	Copper	Austentic stainless steel
Zinc										
Aluminium										
Carbon steel										
Cast iron										
Tin										
Lead										
Brass										
Phosphor bronze										
Copper										
Austentic stainless steel										

Additional corrosion likely under all conditions
Additional corrosion likely under marine exposure
Additional corrosion likely under industrial or marine exposure
Additional corrosion unlikely under all conditions

Figure 4.3

To determine whether or not corrosion is a problem, the performance of both metals must be checked. For example, to check the combination of cast iron with carbon steel, first check cast iron in the vertical columns – this shows no corrosion. However, now check carbon steel in the vertical column and it can be seen that in combination, the carbon steel will corrode in preference to the cast iron. The data does not apply to metals that are immersed in fresh water or seawater.

References

BS PD 6484:1979, *Commentary on corrosion at bimetallic contacts and its alleviation*.

Guides to good practice in corrosion control, National Physical Laboratory (NPL).

Further information

Lifetime Management of Materials Service at the National Physical Laboratory: www.npl.co.uk/lmm

Institute of Corrosion: www.icorr.org

There are also a number of trade and development organisations for metals such as zinc, aluminium, lead, copper, etc., who will provide information on the corrosion performance of metals.

4.2   Building defects

Common defects in buildings
Trevor Rushton

For as long as man has been building there has been building defects but it is fair to say that the last 50 years have seen an unprecedented period of innovation. No longer are traditional craft-based skills passed on from generation to generation; building has become a process of assembly.

Economies brought on by economic necessity, improved design and technology have all tended to make buildings lighter – and now more responsive to thermal and moisture movements. The following table is intended to give a very brief introduction to a number of common or typical materials or construction faults. The list is not exhaustive.

Key to common building types (these are indicative only):

W = Warehouse or industrial
O = Office or commercial developments
H = Housing
A = All types

Table 4.9

Period	Typical problem	Possible effect
	Alterations to trussed (loadbearing) partitions (H)	Removal of support gives rise to distortions in floors, reduction of loadbearing capacity and possible risk of collapse
	Damp penetration through 225mm brick walls (A)	Damage to plaster and finishes, decay to wall plates and bonding timbers
	Defective rainwater goods (A)	Risk of decay in built-in timbers
	Delamination of brick skins (A)	Bulging of brickwork
Pre-1900	Original façade concealed behind later finish of stone, brick or render (A)	Decay of original structure, delamination of skins
	Failed or lack of damp-proof course (A)	Rising dampness, penetrating damp, efflorescence on plaster, decay to skirtings, etc.
	Failure of brick arches and timber lintels (A)	Cracking and distortion of brickwork above window heads
	Insect attack, particularly in poorly ventilated and damp areas such as floor and roof voids (A)	Loss of strength if particularly badly affected

Period	Typical problem	Possible effect
	Lack of restraint to flank walls (O, H)	Bulging or instability, associated cracking on front and rear elevations
	Lead lined parapet gutters – poor outlets and/or poor sizing of lead sheet, leading to splitting and water penetration (A)	Risk of water damage and decay
	Lead water mains (A)	Hazardous to health – partly depends on plumbsolvency (the degree to which the chemical content of local water supplies dissolves lead) of the water
	Over notching of floor joists for retrofit of services (O, H)	Deflection of floors, reduction in loadbearing capability
	Parapet gutters draining to lead secret gutters through roof voids (A)	Risk of blockage and subsequent water leakage/decay. Bird nesting causing blockage.
	Poor quality repairs to roofs and gutters (A)	Risk of timber decay
	Poor ventilation of floor voids (A)	Decay in wall plates, joists, etc.
	Roof covered with concrete interlocking tiles (H)	Overloading of roof structure, bowing of rafters and purlins, roof spread
	Settlement of bay windows (H)	Internal cosmetic damage, distortion in loadbearing elements
	Relaxation of torsional stair treads and landings. (Sometimes termed cantilevered stairs) (A)	Serious loss of structural integrity treads become loose and may fail
	Settlement of internal partitions (H)	Plaster damage, distortion in floors and door openings
	Corrosion of filler joists in concrete floors constructed from concrete containing coke breeze (A)	Regular parallel cracking visible in soffite, circa 600mm centres; can indicate severe loss of section to web of embedded steel beams
1900–1939	Poorly fitting sash windows, risk of decay within window reveals, water penetration beneath sub-sills (A)	Draughty or dangerous operation, decay in concealed areas, lack of security
	Corroded galvanised steel or steel windows (A)	Cracked glazing, high maintenance costs
	Corroded rainwater goods (A)	Risk of decay in built-in timbers, damp penetration

Period	Typical problem	Possible effect
	Corrosion of steel frame (circa 1905 onwards) (Regent St Disease) (A)	Corrosion of frame due to water ingress and lack of effective corrosion protection; leads to cracking of stone or brick cladding following the lines of structural members. Can lead to detatchment of cladding materials.
	Corrosion of roofing nails (A)	Slipping of tiles
	Delamination of cement render finishes to walls (A)	Cracking and bulging of render, detachment of same
	Lead water mains (A)	Possibly hazardous to health
	Outdated electrical services (H)	Possibly dangerous
	Timber joinery (A)	Decay to sills and softwood frames
	Use of boot lintels (A)	Rotation of lintel due to eccentric loading
	Use of mundic concrete (H mainly)	Disintegration and expansion of blockwork leading to failure of rendering
	Wall tie failure in cavity brickwork (A)	Bulging of brickwork, horizontal cracking or 'pagoda effect'
1960s–1980s	Aluminum sash windows (O) A common type of window was the vertical sliding sash. Instead of the vision glazing being held in a frame, the glass ran in aluminium tracks, with horizontal top and bottom frame members clipped onto the glass. Spring balances were used to hold the windows open.	By now, these windows will be very worn. Defects in the springs or breakage of the glass can lead to the ejection of an entire sash window – clearly a health and safety issue. Treat these windows with caution.
	Asbestos (A) Very common in 1960s buildings. Chrysotile (white) for some insulation boards, roof sheets, water tanks, sill boards, etc. Artex, floor tiles, partition wall linings, fire doors, etc. may also contain some asbestos. Amosite (brown) used as insulating boards, fire protection or fire breaks, behind perimeter heaters, partitions, etc. Crocodilite (blue) often paste-applied friable material in boiler rooms, pipework, calorifiers, etc.	Major health risks depending on type, location and risk of disturbance. Deleterious material – detection, management and control are highly regulated. Ask for a copy of the asbestos register. (See also the Asbestos section.)
	Asphalt roofs (O) These could be of quite good quality and may have performed well if laid on a concrete deck.	The lack of insulation would have helped to reduce temperature ranges and so restrict thermal movements.

Period	Typical problem	Possible effect
	Calcium chloride concrete additive (A) Often used by manufacturers of precast elements or for concreting in cold weather. Enables rapid set and removal of moulds. Can also be found in brickwork mortar. Use of unwashed sea-dredged aggregates may have led to chloride contamination, as may exposure to de-icing salt or a marine environment.	Corrosion of reinforcement. (See also Calcium chloride.)
	Calcium silicate bricks (A) Smooth, often creamy coloured bricks made from lime, sand and flint. Small particles of flint can sometimes be seen in cut bricks or weathered surfaces. Can be mistaken for concrete bricks (see page 299). Widespread use in 1960s and 1970s, still manufactured and gaining popularity again. Flank wall of calcium silicate bricks – shrinkage cracking.	Prone to shrinkage (unlike clay bricks which expand after laying). If movement control joints are missed or badly spaced (which they often were), diagonal cracking can occur. Thermal or moisture cracking often visible at changes in the size of panels, e.g. long runs below windows coinciding with short sections between windows. Look out for thin bed cracks and wider cracks to vertical joints. Do not confuse with subsidence cracking or corrosion of steel frame. Its use as a backing to clay brickwork is likely to cause problems as a result of expansion of clay brick and contraction of calcium silicate brick.
	Cladding systems (O, W) Early curtain walling systems relied upon the use of galvanised steel window components coupled together or fixed within framed openings. These systems were often single glazed and incorporated Vitrilite spandrel panels. In the early 1970s more aluminium curtain walling systems were developed. Early systems were single glazed and face-sealed but, latterly, drained systems were installed, incorporating double-glazed units.	Inspect opening sashes for signs of distortion – usually due to paint build up. Window fittings are usually worn or inoperable. Pay particular attention to the security of fanlight fixings. Early double-glazed systems were fully bedded. Deterioration of sealants is common, leading to voiding, leakage and deterioration of edge seals. Be suspicious of early face-sealed systems in terms of future durability.

Period	Typical problem	Possible effect
	Cold bridging and condensation (A) Poor insulation standards led to problems with severe cold bridging, particularly in housing with higher humidity levels. Polystyrene insulation was sometimes used but this was usually no more than 25mm thick. In the mid to late 1970s, thermal insulation standards were increased, particularly in industrial buildings.	Watch out for cold bridging around balcony structures and precast lintels. Provision of insulation within industrial buildings resulted in a spate of condensation problems within roofs. Cold night sky radiation gave rise to condensation on the underside of metal roofing, while poor application of vapour control layers meant that problems were exacerbated.
	Cold flat roof construction (A) Little thought was given to vapour control or, for that matter, roof insulation. It was common to provide sealed flat roof construction with minimal insulation and sometimes a foil-backed plasterboard ceiling lining. Ventilation to the roof void was often ignored. Built-up felt roofs were often asbestos based, but had a life of around 15 years and no more. For this reason most felt roofs would have been replaced by now.	Risk of condensation occurring, with subsequent risk of decay to roof decking or to structure.
	Concrete (A) Can be of mixed quality, sometimes poorly compacted and with lack of cover to steel reinforcement. Under codes, depth of cover for external work should have been circa 40mm.	Sometimes poor durability, corrosion due to the effects of carbonation or chloride content. Tests should be recommended. Poor curing methods could mean lack of durability. Calcium chloride added as an accelerator either in precast or in-situ work.
	Concrete boot lintels (A) Concrete lintels designed to have a projecting nib to support the outer leaf, and built only into the inner leaf, to provide a neat appearance externally.	Rotation of the lintel under eccentric load, creating diagonal cracking to the brickwork above the window. Other signs are opening of the bed joint immediately above lintels and splitting of the reveal brickwork immediately beneath. Once rotation has taken place, brickwork will tend to arch over the opening, thus relieving some of the load on the lintel. Cracks can then be repointed.
	Concrete bricks (A) Similar in appearance to calcium silicate brick but often used in dark brown, dark red or dark grey variants. Harder and coarser texture than calcium silicate bricks.	Suffer from similar shrinkage-related problems. Can be hard to differentiate between these and calcium silicate bricks, but may be harder and contain small particles of visible aggregate.

Period	Typical problem	Possible effect
	Concrete frame (W, O) Expressed concrete frames were common in the 1960s, often with brick infill panels. In the 1970s there was a move away from this form of construction to brick cladding, with the frame concealed either by brick slips or by a brick outer leaf supported on steel angles.	In both cases, there is a risk that the brick panels can become stressed as a result of the normal shrinkage (axial foreshortening) of the concrete frame. A failure to provide movement joints means that loads can be transferred to the panels with the result that the brickwork is disrupted.
	Corrugated 'big six' asbestos cement sheet (W) Often found on industrial buildings and warehouses well into the 1970s. Name given as a result of the 6-inch profile, but in fact 'big six' was one of several different profiles of sheet. Often based on an asbestos content of around 12–15% chrysotile (white asbestos), with profiled eaves and ridge pieces and hook bolt fastenings. By the end of the 1970s insulation was being added to the roof construction and this brought about condensation problems.	Obvious health risks from fibre release. Friable surface, and very fragile – never walk on such a covering without crawling boards. Corrosion of hook bolts will cause sheeting to split. Often coated with bitumen or rubber solutions as a remedial treatment. Be very cautious of the effectiveness of these treatments. Rigid foam spacers were sometimes used below cement fibre sheeting in order to create a void into which insulation could be placed. The rigid foam compresses with time and occasional traffic loads, leading to 'chattering' of the roof sheets and possible water ingress.
	Cracking of brickwork outer skins (A)	Failure to provide for movement resulting from thermal change and/or moisture. Shrinkage of concrete frames coupled with expansion of clay brickwork producing internal stress in the brickwork.
	Cut edge corrosion of Plastisol covered sheet steel roofing (W)	Deterioration of protective covering, corrosion of steel sheet leading to perforation of the base metal.
	External ceramic tiling (O) See 'mosaic tesserae'. Tiles were often prism shaped or ridged in some way.	Similar problems to mosaic tesserae in terms of delamination of background materials.
	Failure of dark coloured Plastisol coatings (W)	Unsuitable colours for roofing lead to high temperature build-up and subsequent deterioration of coating, often large areas of coating flaking away.

Period	Typical problem	Possible effect
	Flat concrete floor slabs (Plate floors) (O, W) Fairly thin slabs with mushroom head thickening around column heads.	Very high shear stress around column head has been found to be cause of structural failure. Beware of flat slab car park construction – e.g. 1997 major collapse of car park of flat slab construction at Pipers Row.
	Glass reinforced concrete (GRC) (A) Lightweight cladding panels, balustrade panels, permanent shutters, planters.	Early forms of GRC contained fibres that deteriorated in alkaline conditions. Loss of strength, cracking and bowing can result. Later alkaline resistant types (Cem-fil) perform better.
	High Alumina Cement (HAC) concrete (O, W) Often used in precast and prestressed work rather than in-situ work, mainly (but not exclusively) for roof and floor beams. X and I profiles were common. Some variants for precast factory units (portals and purlins). Sometimes used to form an in-situ stitch between two precast beam members or column to beam connections. Developed high early strength. Can have a brownish tinge.	Loses strength with age. Susceptible to chemical attack in damp conditions and contact with gypsum plaster. (See also High Alumina Cement concrete [Calcium Aluminate Cement].)
	Hollow clay pot floors (O) Quite common during the 1960s. Concrete poured between pots and in the form of a topping. Sometimes screed could be structural. In other cases, non-structural screeds may have been removed to gain additional load or headroom.	Watch out for clay spacer tiles between the pots. These can conceal honeycombing of the concrete rib, lack of fire protection, durability or strength. Removal of tiles and Gunnite repairs may be necessary.
	Lack of movement control joints (A) Cement masonry walls are prone to thermal and moisture movements. Cement mortar is less flexible than older lime mortars, and the stresses induced by thermal movement are relieved by cracking. Centres of joints depend on nature of brick, size and shape of panel, etc. Lack of joints was common in 1960s buildings.	Oversailing of brickwork on dpc, particularly in warehouses. Also watch out for joints that have been filled with Flexcell impregnated board, as this is not very compressible and can reduce the benefit of a joint. Early sealants were also resinous and could lead to staining of adjoining surfaces. Hardening and embrittlement of joint sealants is to be expected now.

Period	Typical problem	Possible effect
	Large panel buildings (H) A number of different systems were constructed. Large panels formed the external enclosure and also supported precast floor planks or slabs. Connection details were made on site with wire hoops and in-situ work. The design of joints in panel systems was critical in the success or failure of the system. A variety of types were often employed, in some cases using baffles in the form of open drained joints or face sealed joints using mastic or neoprene gaskets. The success of the building from a structural point of view relies on the connection between the individual panels. The ability to withstand local damage by means of alternative load paths is critical.	Risk of disproportionate collapse, e.g. the Ronan Point disaster. Following this, high-rise blocks were strengthened. Some low-rise blocks may not have been checked. Poor quality control of structural connections led to weakness, poor fire stopping or corrosion risk. Possible lack of tying-in of precast components. See 'Tying-in' later in this table. Large panel systems suffer from many of the defects described previously; the more important being rain penetration, corrosion of reinforcement, poor thermal insulation, distortion or physical damage to panels. Calcium chloride was not used in most precast systems, although in some types it was added on a batch-by-batch basis to aid manufacture, perhaps where there were particular programming problems. Thus the inclusion of the material is unpredictable. The most common faults with these types of systems relates to the gradual deterioration of the baffle (particularly where butyl rubber was used), ageing of sealants, or ageing of gaskets in face sealed joints. Misplacement of baffles or insufficiently sized baffles in wide joints can lead to water penetration. It is quite common to find that quality control standards during manufacture were not up to scratch, with the result that reinforcement was commonly misplaced in the fabrication of the concrete panels. This later led to corrosion of the reinforcement and spalling of the concrete. In many cases dry packing used between infill concrete and the panel above is missing or poorly compacted, so that vertical loads are transferred only on the bolt fastenings, resulting in localised cracking around fixing positions.

Period	Typical problem	Possible effect
	Large panel buildings continued	In the long term, panels can distort generally as a result of shrinkage in the concrete structure behind and as a result of normal thermal movements in the building as a whole. This can lead to damage at joints, displacement of seals and baffles and subsequent water penetration. Furthermore, smoke stopping between floors and compartments can be damaged with the result that in a fire, smoke can transfer rapidly between occupancies or zones.
	Lift slab construction (O) A method of multi-storey concrete frame construction where floors are cast on the ground and jacked up into position.	Risk of collapse of structures (e.g. car parks) where corrosion of reinforcement and other factors, such as lack of sheer reinforcement at column/floor connections, gives rise to weakness.
	Mineralite render (O) A thin (2–5mm) coating of fine-grained minerals with a textured surface. Often applied to exposed concrete columns and beams or in larger areas, such as spandrel panels. Variety of colours available.	Beware of adhesion failure – can be widespread. Difficult to match repairs.
	Mosaic tesserae (O) A common finish comprising small (25mm^2) ceramic or glass tiles applied to a render background. Often supplied in paper-backed sheets of around 300 x 300mm to facilitate laying.	Adhesion failure of tiles leads to individual tiles falling off. Render looses adhesion to the concrete or brick substrate. This is potentially more serious as larger and heavier sections could collapse. A hammer survey is to be recommended to check for soundness and to identify hollow areas. Repairs are possible using vacuum injection techniques, but hacking off and repair of spalled areas can lead to peel back and cracking of adjoining surfaces and deterioration due to water ingress and freeze/thaw cycles.
	Mosaic tesserae in overhead situations (O) Often used as a soffit finish to projecting balconies, shopping mall covered ways, etc. Tiles would be bedded on render or possibly applied over expanded metal lathing.	Watch out for corrosion of the metal lathing or fixing screws as these may not have been protected against corrosion. Timber fixing battens behind the lathing can also be a problem. In severe cases, large sections of render and tile finish can collapse. If water penetration is suspected, recommend further intrusive investigation.

Period	Typical problem	Possible effect
	No fines concrete (H) Used in the manufacture of large panels for housing and similar structures, intended to create slightly better insulation properties.	Very low level of resistance to carbonation, hence risk of carbonation and corrosion.
	Panel joints (O, H) Panel joints in large panel systems often comprised a plastic or metal baffle sprung into grooves in the edge of each panel. To prevent leakage, it was common to provide a tape back seal to the rear face of the joint.	Baffles may be missing or dislodged. Back seals often missing with consequent risk of water penetration. Flat roof abutments often dressed under the bottom edge of panels, which makes them very difficult to repair. Often necessary to modify the drained joint into a face-sealed joint.
	Poor cavity tray details (A)	Water ingress
	Poor installation of lateral bracing to trussed rafters (O, H)	Lack of restraint to gables, lateral buckling of trusses.
	Poor quality joinery (A)	Decay of external joinery
	Reconstituted stone (A) Often used as window sills or window surrounds, string courses or other projecting features in all types of buildings. Sometimes fixed with ferrous cramps rather than phosphor bronze. Contain light reinforcement.	Propensity to carbonate fairly rapidly with the result that reinforcement corrodes, causing the features to spall. Corrosion of cramps can lead to displacement of features such as projecting window surrounds.
	Reinforced aerated autoclaved planks (W, O) Often used as roof decks – 'Sipporex' or 'Durox' or sometimes as vertical walling. Thin reinforcement. 300–750mm width. Made from a mixture of cement, blast furnace slag, pfa plus aluminium.	If designed before 1980 may deflect excessively, evidenced by transverse cracking on soffit. Some concerns over durability of reinforcement.
	Reliance on mastic sealant (A)	Poor durability, over-optimistic expectation of longevity.
	Render backgrounds (A) Used in conjunction with tile finishes and mosaic tesserae. Often very strong Portland cement based mixes were used.	Possible adhesion failures on concrete due to presence of traces of mould oil on the surface. Sometimes used water-based bonding agents (giving a white milky appearance when render is removed) when there was a poor mechanical key. Later bonding agents were of SBR, which were more durable. Inflexible renders, high vapour resistance and risk of water entrapment.

Period	Typical problem	Possible effect
	Sand-faced fletton bricks (A) These were a popular and cheap brick type manufactured by the London Brick Company near Peterborough. Often found in 1960s housing or industrial applications. Often, but not always, a red/pink colour with a heavy textured wire cut type of surface. Rear face of brick is smooth with colour bands or 'kiss marks' arising from the burning process.	No problem in sheltered applications but, in exposed situations (such as parapets, chimney stacks and freestanding walls), where saturation is common, bricks are at risk of sulphate attack. The bricks have a very high sulphate content. When wet, soluble sulphates react with Portland cement bedding mortars, causing the mortar to expand and so disrupting the brickwork. Once this occurs, the damage is terminal. Often rendered in the mistaken belief that this will cure the problem, but this is a very short-lived solution and will only make matters worse.
	Softwood joinery (A) External joinery was often of poorly seasoned sapwood with a low life expectancy.	Very poor durability, especially glazing beads, sills and horizontal rails. Further decay where timber has been pieced-in during repair.
	Spontaneous failure of toughened glass (O mainly)	Often caused by nickel sulphide inclusions in toughened glass. Annealed glass not affected.
	Steel windows and cladding (O, W) Typical single glazed windows were manufactured using a section known as W20 by Crittal Windows. Either casements or tilt and turn varieties. Larger curtain walled sections were manufactured by coupling window units together with galvanised steel tee bars.	By now, early windows may be paint bound or distorted. Ironmongery may be defective. Timber subframes were common and may be decayed.
	Stramit roof decking (A) An insulation board often used as a roof decking. It comprises a rigid board about 50mm thickness of compressed straw sandwiched between two layers of building paper. The boards were about 1200 x 450mm width and had a brown paper finish. Check in plant rooms, lift motor rooms, roof access housings, etc. Used in some domestic applications.	The material had a very low resistance to water and would decay easily. For that reason it is less usual to find it nowadays. The boards had a grain and needed to be laid correctly, with support perpendicular to the grain. Failure to do this could lead to distortion of the board and subsequent 'wave' effects in the roof line. This in turn could stress the covering and cause failure. Saturated Stramit board would turn into a brown silage-like mess. Beware of safety issues (risk of collapse) when walking on Stramit roofs.

Period	Typical problem	Possible effect
	System built housing (H) In general terms these types of buildings were prefabricated, based on either steel or concrete construction. Examples would be the British Iron and Steel Federation properties or, if concrete, Woollaway, Unity, Airey, etc. The form of construction was generally based upon the erection of a frame with cladding fitted to it or alternatively a panel system.	Problems have occurred as a result of carbonation in the concrete and initial lack of cover, use of unsatisfactory materials (such as thin steel tube used as reinforcement), problems of interstitial condensation, damage to sealants, etc. In some cases the decay of structural parts has reached severe proportions and it has been necessary to contrive methods of reinforcing the frame or alternatively providing a new cladding system, possibly based on conventional brick and block cavity walling systems.
	Trussed roof construction (H, O) Trusses were introduced into the UK during the mid-1960s and were primarily intended for the housing market, although gradual improvements in stress grading and timber engineering have now taken them into commercial, educational and leisure buildings. Commonly designed to pitches of between 20°–35°, with a span of around 3–10m, trussed rafters are usually jointed with factory-fixed galvanised steel fasteners, although plywood gussets are sometimes used. With the use of stress graded timber, sizes can be reduced to as little as 35mm in width, with trusses arranged at 450mm or more (usually 600mm) centres. Spans of greater than 10m can be achieved, although buckling of compression members can become a problem – and transport to site may be uneconomic.	The correct positioning of the connector plates is essential so that sufficient teeth engage in the timber to prevent the joint, particularly at the apex, from pulling apart. If this happens the truss will settle, or fail. The signs of this may be hogging in the roof and damage to internal finishes. Shrinkage of timber after fabrication can affect the adequacy of the truss as a whole. If the timber members cannot meet, all the joint forces will be taken up by the metal connections, which could buckle or pull out. Corrosion of gang nails used in the manufacture of trussed rafters due to interaction with certain timber preservatives. Compression members may be subject to sideways buckling under load. Bracing is also needed to prevent buckling. Roofs are vulnerable to long-term vibrations which can lead, in an inadequately braced roof, to lateral buckling– either in one or two directions towards gable walls.
	Trussed roof construction continued	The structure will also be expected to afford support to gable walls and this is usually achieved by the use of galvanised steel straps turned down into the cavity and fixed to at least two adjacent trusses. Lack of adequate restraint to gable walls or other unrestrained elements could allow an unacceptable degree of movement to take place.

Period	Typical problem	Possible effect
	Tying-in of precast concrete floor and roof slabs (O, H) Prior to 1972 (CP110) tying-in was left to engineering judgment. There is a need to form a connection between wall structures and internal precast floor planks. This can be achieved with continuous metal straps or structural toppings to prevent planks from gradually moving apart.	Failure to tie-in properly can lead to the elevation gradually parting company from the floors. Evidenced by a series of parallel cracks in the floors, gradually increasing in severity higher up the building. If neglected, collapse could occur under accidental loads.
	Use of brick slips (O)	Risk of slips detaching from the structure owing to shrinkage of the frame and expansion of brick skin.
	Use of built-up-roof coverings (A)	Poor life expectancy of coverings, failure of BS 747 felts due to poor tensile strength.
	Use of deleterious materials such as HAC, chloride, asbestos, woodwool as permanent shuttering (A)	See also Deleterious materials.
	Use of fully bedded glazing methods for thermally insulating glass units (A)	Water leakage, deterioration of edge seals and subsequent misting of glass panels.
	Use of reinforced concrete frames, carbonation and chloride attack (A)	Mechanisms of deterioration in concrete, spalling and corrosion of reinforcement, often compounded by poor cover, poor quality concrete and consequent lack of durability.
	Vitrilite panels (O) Used in conjunction with steel windows (see earlier), these single glazed spandrel panels were made from annealed glass with a powder coating fused into the rear surface during manufacture.	Risk of failure due to heat build-up in spandrel panels, bird strikes or mechanical damage from cradles. Water penetration can cause staining and deterioration of rear surfaces. Replacement panels no longer available and may have been made from painted glass with less life expectancy.

Period	Typical problem	Possible effect
	Wall ties (A) Often wire butterfly ties. Thin steel sections and poor galvanising standards. Cavity walls were rarely insulated and cavity tray detailing may be poor. During the 1960s it was common to use galvanised wire ties and, in some cases, vertical twist ties with substandard protection coatings. The life expectancy of bitumen and zinc coatings on these ties is frequently well under the 60 years that was originally predicted. In fact, in 1981, BS 1243 tripled the minimum allowed zinc coating thickness on wire ties.	Factors which could influence the life of the tie are the steel alloy used, the quality of the protective coating and the mortar type – particularly if this was contaminated with chlorides or if the building was in an exposed location. Research work undertaken by BRE suggests that average zinc loss is about 2.1 microns a year. For pre-1981 ties, this results in a predicted coating life of 12–26 years for wire ties and 25–46 years for vertical twist ties. On the inner leaf, where the circumstances are less aggressive, the zinc coating can be expected to last much longer. If wire ties have been used, they tend to corrode away without any substantial physical disruption to the brickwork. Damage becomes manifest by the sudden collapse of an outer leaf, particularly in conditions of high wind. With the thicker, vertical twist ties the amount of metal is significantly more and if corrosion occurs it is likely that the thickness could increase by as much as four times. The cumulative effect of this corrosion will be the creation of horizontal cracks in the brickwork and eventually the lifting of the roof covering at eaves level to give the so-called pagoda effect.
	Woodwool as permanent shuttering (W, O) Often used in basement car parks where additional insulation was required, or in some office buildings.	Risk of poor compaction of concrete during placing, or grout loss leading to honeycombing around rebars. This could prejudice fire protection, durability or in extreme cases strength. Intrusive investigation required to determine if steel is covered properly.
	Woodwool slab roof decks (A) Often with galvanised steel tongue and grooved edge strips and with a pre-screeded finish, or an applied finish reinforced with chicken wire. Size about 1200 x 450mm or 600mm. (See above for use in permanent shuttering.) The material offered some thermal insulation qualities. Often used in plant room roofs, access housings, etc.	Reasonably durable and, contrary to popular belief, does not degrade rapidly when wet. However, failure of screed is probable during re-roofing operations, leading to need to renew the deck.

Period	Typical problem	Possible effect
1980–present	Thin marble facings (O) New methods of cutting enabled the use of thinner stone as a facing material. Used as a rainscreen cladding or in 'handplaced' situations	Marbles can be prone to curling as a result of isotropic thermal expansion. The effects can be profound, leading to visible distortions and damage to fixings, loosening or cracking of stone and detachment of panels. There is no cure for this type of movement. Calcitic marbles are particularly prone to this form of deterioration.
	Toughened glass, use in overhead situations and in cladding systems (A)	Risk of spontaneous fracture due to nickel sulphide inclusions (see Spontaneous glass fracturing due to nickel sulphide inclusions). Overhead glazing particularly vulnerable in shopping centres, concourses, etc. where failure could result in collapse. Laminated safety glass preferred in these locations.
	Warm edge spacers in insulated glass units (O)	Risk of debonding of the spacer and gradual creeping of the spacer into the sight line.
	External rainscreen systems and external wall insulation systems (A)	Possible defects in fire stopping within cavity leading to rapid fire spread. Risks of detachment or water penetration behind external wall insulation systems, particularly in high-rise buildings where the system may be applied without a drained and ventilated cavity.
	Syphonic drainage systems (W)	A method of water disposal from roofs (particularly to industrial buildings) that relys upon syphonic conditions being created at the outlets and in collector pipes. The system can be very efficient if designed properly, but early systems were not always reliable owing to poor specification, lack of underground drainage capacity and too much time needed for syphonic action to occur. The outlets are vulnerable to blockage and must be well maintained.

Period	Typical problem	Possible effect
	Composite panels (W)	Use of expanded foam insulation in insulated panels and (pre-2000 especially) use of rigid polyurethane foam (PUR) can give rise to total loss in the event of a fire. Fire in a building can spread to the insulation, leading to (in some cases) explosive delamination and rapid fire spread. Polyisocyanurate (PIR) types of foam are now almost universally used for industrial buildings owing to insurance restrictions. Testing of foam type will not necessarily confirm if the panel type is Loss Prevention Standard (LPS) approved but should distinguish between PUR and PIR.
	Liquid roofing systems (A)	Hot applied (generally in new work) and cold applied (for repairs and alterations). Risks of failure due to poor preparation and detailing, contamination of surfaces, etc.
	Indoor air quality (H)	The adoption of improved airtightness standards dictates a need for greater use of MVHE equipment (mechanical ventilation and heat recovery). A failure to service or utilise the equipment properly can lead to a decline in indoor air quality and consequent health risks. IAQ is one of the most significant threats to the health and well-being of building occupants. Poor access to installed MVHE equipment is a factor in problems arising from lack of maintenance; filters become blocked and motors then stall and overheat. Flexible ducts can be kinked during installation resulting in poor performance and a reduction in efficiency.
	Shrinkage and distortion of timber frame (O,H)	Lack of provision for movement coupled with (relatively) static external envelope can lead to distortion in secondary elements such as windows and window frames and damage to internal finishes. Timber frame construction is essentially a process of site assembly of engineered panels or cassettes which dictate very careful erection so as to cater for movement and the transfer of loads to the foundations.

Period	Typical problem	Possible effect
	Poor or inadequate fire stopping /cavity barriers in multi storey (especially timber frame) structures.	Risk of rapid and uncontrolled fire spread leading to significant loss and or risk to life safety.
	Hydrogen embrittlement of high strength steel bolts (W,O)	Entrapment of hydrogen gas during the manufacturing process (often by coating) leads to the gas being absorbed into the metal and concentrating around areas of high stress; can lead to sudden failure of the bolt or component. Tends to affect high strength steels more than lower grade metals.
	Decay in timber floor or roof cassettes (A)	Can be a result of poor vapour control and lack of effective ventilation; often caused as a result of a failure to dry out a structure adequately following unintended saturation during construction and prior to the installation of permanent weathering.
	Poorly installed external wall insulation systems (A)	Can affect all building types but mainly housing. EWI systems can be effective if installed properly but are often poorly installed. On tall buildings they need to be treated as a rainscreen cladding and so have adequate means of cavity drainage and fire stopping. Such measures are often neglected leading to water ingress. Adhesive fixing of insulation slabs needs to be controlled carefully failing which delamination under wind exposure can occur. Mechanical fixing of slabs is preferable. Polystyrene insulation materials perform poorly in respect of moisture transfer and can lead to the entrapment of moisture behind linings.
	Timber rainscreens (A)	Often applied as part of a low or zero carbon initiative. Commonly, defects with breather membranes and poor cavity tray detailing leading to uncontrolled water entry and passage behind the cladding and subsequent water ingress into the building. Moss and algae growth in poorly detailed joints leads to water retention, staining and gradual deterioration.

Period	Typical problem	Possible effect
All periods	Over notching of floor joists (W, O, H, A)	Reduction in strength, sagging
	Removal of chimney breasts (W, O, H, A)	Possible lack of support
	Provision of insulation, blocking ventilation paths (W, O, H, A)	Condensation
	Blocking of airbricks (W, O, H, A)	Lack of ventilation, risk of decay
	Removal of loadbearing walls or walls affording stability (W, O, H, A)	Possible long-term structural consequences
	Removal of, or planting of, trees or large shrubs (W, O, H, A)	Possible desiccation or rehydration of subsoil, damage to drains or foundations
	Replacement windows (W, O, H, A)	Poor support to bay windows, distortion of brickwork. Poor sealing of new windows to existing fabric leading to loss of airtightness and water penetration.

Steel frame corrosion (Regent Street disease)
Trevor Rushton

Until the early part of the 20th century, substantial buildings tended to be constructed using load-bearing masonry – at least for the external walls. Because of the massive nature of these forms of construction, the walls were able to accommodate movements due to temperature and moisture, as well as small building movements, without significant harm. With the advent of steel as a versatile construction material, the position changed dramatically. By 1910, steel framing was becoming very popular, and this meant that walls could be reduced in thickness, with obvious benefits in terms of economy of material, weight and cost. The use of these construction methods in London's Regent Street has given rise to the common description of Regent Street disease, but it is a problem that is by no means confined to this location.

Typically of many new construction materials, the properties of steel were not fully understood, or at least if understood, ignored. From the early part of the 20th century up to the Second World War, it became common practice to construct load-bearing frames of steel, clad on the external faces with stone, brick, or terracotta. The external cladding would be notched around the steel frame, with the void between the two filled with low-grade mortar.

As we now know, moisture and oxygen can permit the formation of corrosion cells which can cause significant delamination and loss of strength. The main problem is that corroded steel has the propensity to expand to at least 4 times its original volume (some sources suggest up to 10 times). Given that the voids around the steel were filled, the expansion of the steel would inevitably result in cracking of the stone, and in extreme cases the loosening or loss of support to horizontal stonework, which could then collapse. Such problems could occur long before the corrosion has compromised the ability of the structural member to perform properly.

While some corrosion protection was common after about 1930, the methods used were unlikely to offer a long-term benefit, and so it is now very common to find evidence of corrosion in many steel-framed, brick, or stone-clad buildings.

Diagnosis

Evidence will take the form of vertical or horizontal cracks reflecting the location of the steel frame. Parallel cracks indicating a column position may be less serious than horizontal cracks to a beam location (as these could indicate that the stone is in danger of falling away), but nevertheless investigations are needed to determine the condition of the steel. While early evidence of corrosion is reflected by hairline cracks, more significant problems (particularly in glazed brickwork) can indicate advanced corrosion or loss of section.

Accurate diagnosis is usually reliant upon exposure of the element concerned, but non-destructive methods such as Resistivity Metering, dynamic impedance, or ultrasonic testing are sometimes possible.

Dealing with corrosion using traditional methods is expensive and disruptive. The steel must be exposed, cleaned and protected – not an attractive proposition when dealing with a listed building or an important facade. It is usual to provide a corrosion barrier to the steel and then to create a void around it so that if further corrosion does occur, it will not result in cracking.

Because of the cost and disturbance of these forms of treatment, more attention is now being paid to the application of impressed current cathodic protection systems, which rely upon the concealment of discrete anodes into the stone joints, and electrical connection to the steel frame and the introduction of an electric current to reverse the corrosion current. These systems require very careful design and installation and it is imperative that the entire frame is protected in this way to prevent stray currents from having a harmful effect.

Further information

Gibbs, P., *Cathodic protection of early steel framed buildings*, Monograph No. 7, Corrosion Prevention Association.
Corrosion in masonry clad early 20th century steel framed buildings, Technical Advice Note 20, Historic Scotland Technical Conservation Research and Education Division.
Warland, E. G., *Modern practical masonry*, London, Sir Isaac Pitman & Sons Ltd, 1929.
Rushton, T., *Investigating deleterious and hazardous building materials*, RICS Books.

Fungi and timber infestation in the UK
Fungi and moulds
Trevor Rushton

Mould growths and wood-rotting fungi are possibly the most common defects that will be encountered in buildings, particularly when inspecting domestic accommodation. Mould and fungi are effectively base plant forms, the moulds draw nourishment from air and fungi from within host materials.

Wood-rotting fungi are familiarly categorised as wet or dry rot. While there are numerous forms of wet rot, there is a single true dry rot, Latin name *Serpula Lacrymans*.

Fungi require both a source of nutriment and appropriate levels of moisture. Dry rot characteristically requires a lower average moisture content within the host timber than wet rots, though to some extent dry rot is able to manipulate its environment to create more favourable conditions. Dry rot is known to grow through plaster and masonry in a search of fresh sources of nutriment, but takes no nutriment from the plaster or masonry itself.

The *Fungi identification table* below provides an overview of the most commonly encountered wood-rotting fungi in the UK. Hyphae (thin strands often mistaken for cobwebs, that make up the mycelium) spread out from the germinated spores, however it is often the sporophores (fruiting bodies) that are the first recognised indication of an infestation.

Numerous companies offer rot treatment works within the UK, many being members of the British Wood Preserving and Damp-proofing Association (BWPDA). However, beware of recommending so-called 'specialist surveys' as the specialists will have a vested interest in any repair recommendations that they may make.

Traditionally, treatment of dry rot included:

- stripping off of plaster to 1m beyond the last identified point of infection;
- removal of visible indications of the fungi;
- cutting back of affected timbers to 500mm beyond the last recognised point of infection;
- surface spraying of an approved fungicide; and
- irrigation of the surrounding walling.

Any new timber introduced to the area needs to be pre-treated in accordance with BS 5268-5:1989, remaining timbers being cleaned and then treated, to BS 5707:1997, with a solvent preservative, and possibly with application of a preservative paste.

In recent years irrigation has come to be considered of limited benefit, not least because of the amounts of liquid required and the potential for resultant damage. It is currently considered more important to reduce the moisture ingress in the area concerned and promote rapid drying in order to deprive the fungus of one of its essential life sources. Any remaining fungus within the masonry should then remain dormant, even if it does not die off completely. Such forms of treatment are to be commended, particularly where historic buildings are concerned. Wet rots are potentially less destructive and easier to treat than dry rots, though again stopping the source of the moisture should be a significant concern.

With regard to remedial works, developments in resin systems in recent years have allowed less intrusive repairs and a reduction in the need for significant timber replacement. This is particularly beneficial when dealing with historic buildings or significant structural timbers. However the works are undertaken, the assurance of an insurance-backed guarantee should be sought rather than reliance on the treatment company's own certificate.

On a cautionary note, remember that the treatment systems utilise powerful chemicals. Safety procedures must be followed and current safety legislation must be adhered to, in particular the Control of Pesticides Regulations 1986 (COPR), as amended in 1997, and the Control of Substances Hazardous to Health Regulations 2002 (COSHH), as amended.

Consideration must be given to any wildlife that may come into contact with the works. Under the Wildlife and Countryside Act 1981, the Wildlife and Countryside (Service of Notices) Act 1985, and the [European Community] Conservation (Natural Habitats, etc.) Regulations 1994 (as amended in 2007), treatment may require approval by one of the conservation agencies forming the Joint Nature Conservation Committee (JNCC), or indeed other appointed agencies.

Moulds are an increasing problem within the carefully controlled atmospheres of modern buildings. The advent of double glazing systems, and other modern building techniques, has often restricted air flow within properties and resulted in ideal environments for these microscopic plant forms to grow.

Table 4.10 Fungi identification table

Type	Usually found	Effect on timber	Mycelium	Fruiting body	Conditions for growth
Wood rotting fungi *Dry rot*					
Serpula Lacrymens	Inside buildings, mines, boats – never attacks timber outside	Large cuboidal cracking (brown rot)	Cotton-wool-like if damp; greyish white with purple/yellow and lilac patches if dry	Reddish brown centre, white margins; flat plate or bracket shape; possibly red spore dust nearby	Timber MC 20–40% (slightly damp) Temperature 0–26°C
Wet rots					
Coniphora Puteana (cellar fungus)	Most common of wet rots in buildings; associated with serious leaks – failed plumbing, etc.; also decays exterior	Cuboidal cracking – small cubes (brown rot); may leave thin veneer of sound timber; affected wood becomes dark brown	Brown branching strands on wood and masonry or brickwork; usually not in daylight areas	Rarely found inside; flat plate-like; greenish brown centre, yellow margin; knobbly surface	Timber MC 45–60% (very damp) Temperature –30°C to +40°C

Species	Association / location	Damage to wood	Mycelium / strands	Fruiting body	Conditions
Fibrioporia Vaillanti (mine or pore fungus)	Associated with water leaks; most common species of poria group	Cuboidal cracking – large cubes (brown rot); affected wood darkens	Strands flexible when dry; white	Plate-shaped; white pores; rare	Timber MC 45–60% (very damp) Temperature up to 35°C
Phillinus Contiguous	Decay of external joinery (softwood)	Timber becomes soft (a white rot). Wood becomes fibrous	Light brown masses	Plate-like with pores; dull brown	Timber MC 22%+ Temperature 0–31°C
Phillinus Megaloporous	Attacks oak heartwood; presence often associated with death-watch beetle		Yellow	Large, plate-like, hard; various browns in colour	Timber MC 20–35% Temperature 20–35°C
Corioius Versicolor (Polystictus)	Most common white rot decay of external hardwood	No splitting or decay but much weight loss	Rarely seen	Up to 25mm across; hairy ringed zones to pores to underside	
Lentinus Lepideus (Stag's Horn fungus)	Rare, but sometimes in flat roofs	Cuboidal cracking Darkens woods; wood feels sticky	Soft whitish needle- shaped crystals on surface	Some resemble stags horns, others are inverted mushrooms on stalk; brown	Timber MC 26–44% Temperature 25–37°C

Type	Usually found	Effect on timber	Mycelium	Fruiting body	Conditions for growth
Non-wood rotting fungi					
Peziza (Elf-Cup)	Occurs on saturated masonry or plaster, internally and externally; associated with leaks			Buff coloured and fleshy; distinctive	
Moulds					
Aspergillus Penicillium Pullularia	Almost any damp surface in humid conditions	Superficial – easily removed	Like coconut matting	Toadstool; white head; spores released in black ink type liquid Microscopic but spores show up as various colours: black, green, white, brown, yellow, pink	Very humid conditions
Strachybotrys Chartarum (Toxic mould)	On high cellulose content materials	Superficial to timber – various possible health risks	Green or black spots/mats		Elevated humidity

Mould has been a significant concern in the US for some time, with black mould, *Strachybotrys Chartarum*, being blamed for various symptoms, including aggravation of asthma and rashes. Site operatives may display flu-like symptoms, as a result of Organic Dust Toxic Syndrome, when working in areas where widespread fungal contamination exists.

More seriously there is some evidence to suggest that black mould may also lead to bleeding in the lungs of infants exposed over long periods, and in turn to pulmonary hemosiderosis – a lung disorder in which bleeding (haemorrhaging) into the lungs leads to an abnormal accumulation of iron.

As well as reduction in moisture levels, improved ventilation is important in the fight against mould growth. Various proprietary chemical mould treatments are available and will help in the short term, although treatment of the source is far more effective. As with rot treatment, appropriate precautions and legislation will need to be considered.

Further reading

Digest 299, *Dry rot: Its recognition and control*, Building Research Establishment (BRE), 1993.
Defect Action Sheet 103, *Wood floors: Reducing risk of recurrent dry rot*, BRE, 1987.
Good Repair Guide 12, *Wood rot: Assessing and treating decay*, BRE, 2013.
Report 453, *Recognising wood rot and insect damage in buildings* (3rd edition) (2010 reprint), BRE, 2003.
T1/595/2, *Fungal decay in buildings*, Wood Protection Association, 2013.
Information Paper 11/85, *Mould and its control*, BRE, 1985
Digest 370, *Control of lichens, moulds and similar growths*, BRE, 1992.

Insect infestation
Trevor Rushton

Within the UK, wood-boring beetles are the major group of timber attacking insects. Wood-boring beetles are often erroneously referred to as 'woodworm', possibly because it is the larval stage of the insect's life cycle that eats into the wood before emerging as adults to start the breeding process over again.

The female beetles lay eggs within cracks or end grain of timber, and from there the eggs hatch and feed on the nutriments contained within the wood, leaving frass (digested waste) in the tunnels they form. The maturation can take up to 11 years (depending on the species of insect) before the adult beetle emerges from the timber. The common furniture beetle (Anobium Punctatum) has a life cycle of 3 years, and is the most commonly encountered of the beetles in the UK. The *Beetle identification table* below details the wood-boring beetles commonly found within the UK.

It is often the adult beetle's flight holes that are the first indication of infestation, by which time the damage has been done. The extent of damage will however depend on the type of beetle and the number of life cycles that have occurred before the damage has been observed.

Treatment with low odour insecticidal fluids can take the form of spray or brush application or gel or injection treatments. Fumigation may also be considered in certain buildings, though it is less effective as it does not directly target the infestation.

As with rot and mould, treatments utilise strong chemicals and the correct safety procedures and legislation must be adhered to (see the previous pages on fungi and moulds).

Termite problems in Europe are increasing and have been reported in the UK, though as yet not in significant quantities. Broadly speaking, treatment and eradication methods are similar to those for wood-boring beetles, although there are some significant differences. Unlike our common wood-boring beetles, termites are social insects living in colonies. The time span for damage caused by an infestation can therefore be significantly less than that of the more widely encountered wood-borers.

Table 4.11 Beetle identification table

Species	Identification	Flight hole	Adult (not actual size)	Grub (not actual size)
Common furniture beetle	Very common, estimated that up to 80% of houses over 40 years old in rural areas are affected. Infestation often in damp areas of house, for example, beneath WC. Beetles are around 3mm long. Adult beetles emerge May to September.	1.5–2.0mm diameter		
Death-watch beetle	Infestation uncommon, often found in ancient buildings and therefore more expensive to eradicate. Confined to south and central parts of England and Wales. Attacks elm, chestnut and oak. Presence may indicate fungal attack. Adult beetles up to 8mm long emerge in spring.	Up to 4mm wide		

Species	Identification	Flight hole	Adult (not actual size)	Grub (not actual size)
Waney edge borer	Found in timber where bark not completely removed. Larvae confined to bark areas and hence damage caused is superficial.			
House longhorn beetle	Only found in Surrey, Berkshire and Hampshire. Beetles are up to 25mm long. Regulations now require new timber to be treated prior to use. This is a very large borer.	Oval flight holes up to 9mm x 6mm		
Forest longhorn beetle	Will attack softwoods and hardwoods in freshly felled lumber or standing trees; will not attack seasoned wood. Life cycle varies according to species. Treatment is usually unnecessary.	Oval, up to 10mm across. May be plugged with coarse fibres.		
Wood-boring weevils	Several species – only attack partially decayed timber, cause considerable damage. Beetles are 3.5mm long.	1mm		

Species	Identification	Flight hole	Adult (not actual size)	Grub (not actual size)
Termites	Varying in size from 4–15mm long and in colour from white to tanand black, termite infestations are usually obvious by the presence of characteristic dry pellets in wood or on horizontal surfaces beneath infested wood. Darkening or blistering of wood in structures is another indication of an infestation.	·		n/a

Drawings contributed by Fiona Mackay and Stephen Rickerd

Condensation
Trevor Rushton

What is condensation?

In its gaseous form, water exists in air as water vapour. Air itself is a mixture of gases of which water vapour is just one; air does not carry water as such. Water vapour exerts a pressure; when we talk about partial pressure we mean that part of the total gas pressure formed by water vapour. The warmer the air the greater the percentage of moisture vapour it can support.

The amount of water vapour in the air depends upon its temperature and pressure. The maximum amount of water that can be held at a given temperature occurs when air is said to be saturated. At saturation point as much water is being lost from a body of water as is returning to it. A measure of the amount of water vapour is 'Relative Humidity'

(Rh) which is expressed as a percentage of actual vapour against saturated air at the same temperature. 100% Rh means that the air is saturated.

For a given amount of water in a given amount of air, there is a temperature at which the air will become saturated. The temperature at this point is called the 'dew point'. It follows therefore that for condensation to form on a surface, that surface must be at or below the dew point temperature.

Freely available psychrometric charts enable one to determine the relationship between the moisture content of air, the temperature, the partial vapour pressure and the dew point to seek to determine whether condensation is occurring.

An average household will generate about 12 litres of moisture per day through bathing, washing clothing and cooking. If paraffin or bottled gas heaters are used this will rise to over 20 litres a day, and will increase further if clothes are dried indoors. Additionally, each person exhales approximately 1 litre of moisture per 24 hours.

Rh is critically important to indoor air quality. Moulds and spores that are above 80% Rh can be propagated; it is important to remember therefore that critical conditions can exist before liquid water (condensation) forms on a surface. Wherever possible, internal relative humidity should be kept below about 60%.

In kitchens and bathrooms on inherently cooler tiled and ceramic/enamel surfaces, condensed-out moisture can be wiped away and does not usually offer a place where atmospheric moulds and fungal spores can propagate.

On slightly warmer surfaces, such as papered walls and ceilings and woodwork, the mould and fungal spores have a far greater chance of propagating and growing to form both toxic and non-toxic growths that can both discolour decorations and, through mycotoxins, affect the health of inhabitants.

In modern, highly insulated and airtight homes, condensation risk must be mitigated by good design and by maintaining an efficient mechanical extraction and heat recovery system. Given the probability that these systems will not be maintained properly during their life and that users may be ignorant of the correct operating methods, there exists a significant risk that indoor air quality (of which condensation is a major factor) will deteriorate, leading to various health-related problems and damage to the building fabric.

Mechanisms for reducing condensation include:

- Have adequate natural permanent ventilation and temporarily ventilate rooms (to the outside) after cooking or bathing, or whenever you see condensation forming on cold surfaces such as glass to windows and doors.
- Avoid drying washing indoors or do so only in rooms with open windows and closed internal doors.
- Avoid using flueless gas and oil, especially paraffin heaters.
- Maintain natural 'wind tower' air movement paths in habitable buildings.
- Rooms that are mechanically ventilated (such as bathrooms and toilets) should be fitted with humidistat-controlled extract fans.
- In bedrooms, ensure that (at the very least) there are trickle ventilators to remove exhaled water vapour from sleeping persons.
- Maintain adequate working and operable windows to all habitable rooms.
- Improve the overall level of insulation to the property and install appropriate double- or triple-glazed openable window systems.
- Provide adequate background central heating, especially to bedrooms, but without doubt to all habitable rooms, to bring the overall building up to a general fabric temperature that will mitigate against condensation formation. There should be low levels of background heating throughout the day and night in cold weather

even when no one is at home. It is important not to rely on sudden bursts of high heating levels for short periods. Ideally a thermostatically controlled system should be in place to maintain the habitable rooms at a minimum air temperature of 10°C with the ability to rapidly increase individual rooms or the whole system to between 17–21°C when the room or property is occupied.

See *Toxic moulds* for some of the negative consequences of condensation.

Further information

Condensation risk – impact of improvements to Part L and robust details on Part C, Final report: BD2414 DCLG 2011.
BRE Information Paper 02/05, *Modelling and controlling interstitial condensation in buildings*, 2005.
BS 5250:2011, *Code of practice for control of condensation in buildings Year 2011*.

Toxic mould
Trevor Rushton

Concerns over health issues arising from exposure to mould growth are becoming widespread, fuelled partly by media reports and misinformation. At the present time there is a general lack of accurate and reliable information on specific human responses to mould contamination and no proven methods of measurement to establish the degree of exposure needed to trigger symptoms associated with mould exposure. The issue is confused because mould spores are present everywhere in the environment, and responses to exposure vary widely between individuals. Some people suffer no apparent symptoms, whilst atopic or sensitised individuals can experience negative health effects very shortly after exposure.

There are many thousands of different moulds and they all occur naturally in our environment. Health risks arise primarily as a result of inhalation of the spores or fragments of spores; it is not so much exposure to normal levels of spores or fragments as exposure to excessive levels which may trigger asthma, cause allergic illness, respiratory infections, or may bring about toxic effects from certain chemicals in the mould cells.

Potential human contact can arise as a result of the inhalation of any or a combination of:

- the living mould itself – the live plant;
- the spore – spores can lie in a dormant state for a long time waiting for the right conditions for growth; dead material; protein and other molecules that are present even when the mould is no longer alive incapable of growing; and
- mycotoxins – the waste products generated by moulds.

Disturbance of moulds or fragments can lead to high concentrations and thus greater risk; for this reason it is often the case that comparison samples are taken between internal and external environments to determine whether levels are elevated; the difficulty being that with no established and agreed measurement data it is difficult to know what would constitute a trigger level for action to be taken.

Propagation

Fungal mould spores in the air will develop when they alight on parts of the building that provide the right substrate and conditions for growth.

Key requirements for propagation are:

- a source of infection
- oxygen (although some moulds are obligate anaerobes, which means they can grow in low concentrations of oxygen (conversely, obligate aerobes require oxygen.)

- moisture
- a suitable temperature (some moulds flourish in very warm temperatures, others prefer lower temperatures. Very few, will grow in cool or cold temperatures.)
- food source

Different environmental conditions can facilitate the growth of different organisms; for example, very damp or wet conditions in the presence of plasterboard are an ideal environment for the growth of Srachybotrys. Damp, warm conditions in combination with dust and dirt can, over a lengthy time provide ideal conditions for Penicillium.

Key moulds in the UK are:

> *Strachybotrys chartarum* – a black or greenish-black mould that grows on material with a high cellulose content, including building materials such as the paper facings of plasterboard or chip and particle boards when these materials become water damaged. This mould requires very wet or high humidity conditions for days or weeks in order to grow.
> *Aspergillus* – this comprises a family of some 185 subspecies, of which some 20 or so are presently considered dangerous, including fumigatus, flavus, and niger. It is usually black, green, or grey, although other light colours are known. Excessive indoor humidity from water vapour condensing on walls, plumbing failures, splashes from bathing or taking showers, or water ingress from outside may lead to the growth of many mould varieties, including *Aspergillus*, *Blastomyces*, *Coccidioides*, *Cryptococcus*, and *Histoplasma*, as well as *Strachybotrys*.

Objectives of investigation

Obvious mould growth needs to be attended to; testing and swabbing can be undertaken by specialists with the following objectives:

- To determine if the building materials or contents are colonised by mould
- To determine whether the indoor environment has 'normal and typical' types and amount of airborne mould spores
- To quantify and identify fungal biomass and determine if levels and types of fungi are at variance with outdoor level due to fungal growth
- To determine if specific disease-causing agents such as *Aspergillus* or *Fusarium* are present in significant amounts
- To determine presence of viable and non-viable spores, hyphal and micro fragments
- To assess MVOCs (Microbial Volatile organic compounds) to assist in the identification of hidden mould sources
- To reduce exposure to occupants who may be atopic or susceptible to mould

Methods of sampling

- Mycometer – using swabs and air samples to collect contamination for quantification of levels of biomass
- Air-o-cell cassettes for rapid collection of samples for laboratory analysis – not always possible to separate specific types of spore
- Lift tape or Bio tape – enables assessment of whether moulds are part of the general fallout from the exterior environment or whether active moulds are present
- Gas chromatography – non-destructive testing for VOCs and MVOCs to determine the existence of hidden mould

Remedial works

The best option for buildings and their occupants is to ensure that conditions do not exist whereby moulds of any variants can propagate. This is best achieved by keeping the buildings warm, dry, and free from external water ingress into the fabric of the building. Persistent surface condensation is one of the normal triggers for growth although it is not the only one. Dealing with the predominant cause is the best way of treating the problem – this may involve a change in occupant behaviour patterns as well as changes to the building fabric or services. Treating moulds with biocides is not to be commended – for one thing, the chemicals could be harmful, and for another, the decayed residue of moulds can be as problematic for sensitive individuals as the live mould or spore itself.

See also *Condensation*.

Further information

BRE publications DG85 and DG297 (both 1985), DG370 (1992), and IP12/95 (1995) are all worthy of consultation.

Toxic moulds and indoor air quality, Jagjit Singh of Environmental Building Solutions, Ltd describes the relationship between toxic mould and indoor air quality. Available at www.ebssurvey.co.uk/docs/Toxic%20 Moulds%20%20indoor%20Air%20Quality.pdf (Indoor Built Environment Review Paper, 2005, 14;3–4:229–234, accepted 21 February 2005).

Research into mould and the implications for chartered surveyors, RICS, November 2004, has most useful and thought-provoking lines of enquiry and research data into such moulds.

WHO guidelines for indoor air quality: dampness and mould, published by the World Health Organisation Regional Office for Europe in July 2009 (ISBN 978 92 890 4168 3) is a most useful document and contains what are effectively their considered guidelines for the protection of public health from health risks due to dampness, associated microbial growth, and contamination of indoor spaces. Available at: www.euro.who.int/document/ E92645.pdf

The City of New York has an excellent website which gives useful guidance and advice; see www.nyc.gov/html/doh/ downloads/pdf/epi/epi-mold-guidelines.pdf

The Australian websites www.abis.com.au/toxic-mould and www.mould.com.au

Rising damp
Trevor Rushton

'Rising damp' is the process by which moisture rises vertically up a wall as a result of capillary action. Many will argue that rising damp can only exist if the source of the moisture is contact with damp soil. In many cases, examination of facts will reveal that the source of damp is something other than simply 'rising damp' from the ground. Penetrating damp from high external ground levels or leaking service pipes, for example, may produce similar visual symptoms.

Rising dampness within a wall is in a sensitive equilibrium. There must be a supply of water at the base of a wall and the height to which that water will rise depends upon the pore structure, the brick, plaster, or other finish. Water will also evaporate from the surface of the wall at a rate dependent upon temperature and humidity.

During wet weather, evaporation may decrease and groundwater tables may rise, giving rise to an increase in the severity of the dampness. The reverse may happen during dry spells, and evaporation will be increased by central heating.

Soluble salts are present in many building materials. The salts can be dissolved and moved to the surface of the element as evaporation takes place. Hygroscopic salts (typically, nitrates and chlorides from groundwater) can be present in some materials. These salts absorb moisture from the atmosphere, and can in certain circumstance cause extensive staining and disruption of finishes.

Salts will increase the surface tension of the water and so draw it farther up a wall. Furthermore, as evaporation occurs, stronger salt solutions are drawn towards the surface and may eventually crystallise out. This process reduces the amount of evaporation and so may raise the height of the dampness. The soluble salts are often hygroscopic and absorb moisture from the atmosphere. If this occurs, the situation will appear worse during wet weather and better during dry.

As noted above, the presence of hygroscopic salts does not necessarily indicate rising dampness.

Research by the Building Research Establishment (BRE) and others suggests that rising dampness is often misdiagnosed by surveyors and so-called damp specialists with the result that costly and unnecessary remedial treatments are specified.

In many cases, diagnosis is undertaken by means of electrical resistance or capacitance meters, but these can give very misleading and unreliable results in materials other than timber. It is fair to say that if the meter reveals the wall to be dry, then it probably is dry; the problem comes when a meter records something as damp. Surveyors should not diagnose rising damp without first having undertaken a proper study.

When rising dampness is suspected, do not automatically recommend a specialist inspection – more often than not the specialist will use exactly the same resistance equipment to make his or her diagnosis. A more reliable method has been prepared by the BRE and may be found in BRE Digest 245, 2007.

Symptoms

Typically symptoms may include:

- damp patches;
- peeling and blistering of wall finishes;
- a tide mark 1m or so above floor level;
- sulphate action;
- corrosion of metals, for example, edge beads;
- musty smells;

- condensation; and
- rotting of timber.

The above symptoms do not of themselves indicate the cause of dampness. Common causes could be lateral rain penetration, condensation, or entrapped moisture; hygroscopic salts such as nitrates and chlorides may also reveal themselves as damp patches. High external ground levels, bridging of damp-proof courses, defective rainwater goods, and the like should all be self-evident and could give rise to similar symptoms. Other possible sources of salt contamination include chemical spillage, splashing from road salt, etc.

Equilibrium moisture content

Many building materials absorb moisture, and, when exposed to damp air, will attain equilibrium moisture content. In this case we are dealing with water in its gaseous phase, i.e. water vapour. Moisture absorbed via exposure to water vapour is termed 'hygroscopic moisture'; the amount of water present from this source is termed 'hygroscopic moisture content' (HMC). Although the response to change is not immediate, the amount of hygroscopic moisture will vary according to the moisture content of the air (expressed by vapour pressure) and will eventually settle at an equilibrium state. This equilibrium condition will vary according to relative humidity. Typical relationship curves can be plotted for different materials, and, although these can only establish general indications, it is possible to compare readings from different materials in the construction of a wall. For example, at 75% relative humidity (Rh) the moisture content (MC) of yellow pine would be 13%, while the MC of lime mortar would be 2% and 0.5% in brick.

By using relationship curves and determining the Rh, one can gain a rapid appreciation of whether the moisture content is greater than the equilibrium moisture (EM) content – if it is greater than the EM it follows that moisture must be coming from a source other than the atmosphere.

Some materials possess an HMC of as much as 5% without the introduction of salts from external sources. This figure should be regarded as an appropriate threshold as to whether or not remedial action is likely to be required.

Measurement of moisture content

Resistance or capacitance meters can give misleading results and must be used with care to give an initial indication.

The presence of soluble salts on the surface of a wall will cause an electrical resistance meter to indicate a high reading, even if the wall were otherwise dry. Deep wall probes may give a more accurate picture, but will still be affected by soluble salts, as these are generally highly conductive.

A Speedy Moisture Meter (or carbide meter) will give a much more accurate reading of MC in all materials. (Resistance meters are usually calibrated for use in timber and can give an approximation of MC in that material.)

The Speedy meter comprises an aluminum flask fitted with a pressure gauge and a removable lid. Using a 9mm drill on slow speed, a sample of dust is taken from the brick, mortar, or plaster. The sample is weighed and placed into the flask. A small quantity of carbide is then added and the flask sealed. Moisture in the sample reacts with the carbide to form acetylene gas. The pressure of that gas is then read off the pressure gauge, which is calibrated to read %MC. With care, the meter can give a very accurate reading, comparable with laboratory kiln dried tests. For brick and masonry, readings above 2% should be investigated, although the threshold can vary according to the nature of the material being tested.

Diagnosis of rising damp

BRE Digest 245 sets out a method of diagnosis. The method involves drilling samples from the wall to measure both their moisture content (MC) and hygroscopic moisture content (HMC). Samples are taken from mortar joints from 10mm to a depth of 80mm every two or three courses from floor level up to a level beyond that which damp is suspected. While the carbide meter can be used to measure MC it will be necessary to send samples to a laboratory to measure HMC – to see if the samples could have absorbed the quantity of moisture found from the atmosphere.

By subtracting HMC from total MC, it is possible to determine the value of 'excess' moisture, which could result from capillary action or water from other sources. The comparison of HMC and MC gives an indication as to which is controlling the dampness at any position. If MC is greater than HMC, then moisture is coming from some other source. If the reverse applies, then moisture is coming from the air. Plotting the results graphically can then assist in gaining an accurate picture of what is happening.

The process of diagnosis of rising damp is time consuming and should only be undertaken once all possible causes of dampness have been eliminated.

Surface damage arising from hygroscopic salts can be significant. The HMC of contaminated wallpaper or plaster can be as much as 20%. It follows, therefore, that contaminated plaster will need to be removed. (BS 6576: 2005 deals with this subject in more detail.)

Treatment of rising damp

Assuming that the source of the dampness is confirmed as rising as opposed to penetrating dampness, the most likely solution will be to insert a new damp-proof course (dpc), either by physical insertion or by chemical injection.

Physical insertion will provide a reliable barrier and will be appropriate in conditions where total certainty is required and possibly where it is necessary to extend an inner damp-proof membrane up to the dpc level and to make a physical connection with it. Care is needed during the insertion process, both from a health and safety point of view (the method uses a carbide tipped chainsaw) but also to ensure that the formed slot is properly wedged and packed upon completion to prevent settlement cracking.

Physical insertion is rarely undertaken because of the risks involved. Chemical insertion is used widely; there are a variety of aqueous and solvent-based systems, as well as injection mortars and thixotropic systems. Such treatments may be introduced using pressure or gravity. The BRE do not favour the use of alternative treatment systems such as siphons or electro-osmosis as these can be unreliable.

Syphon tubes are claimed to be an effective, environmentally friendly method of treatment. However, the systems rely upon the evaporation of moisture via (usually) a ceramic liner – the effectiveness of the tubes decreases with age as soluble salts are deposited at the air boundary.

In certain circumstances, improvements can be made by simply reducing ground levels and/or by controlling external water systems. The provision of perimeter land drainage may be more appropriate for older or listed buildings, where physical works may not be desirable or even practicable.

Replastering is often necessary, as the existing plaster finishes may be contaminated with salt or in a deteriorated condition. In ideal circumstances, use plaster that is highly vapour-permeable while at the same time able to act as a barrier to hygroscopic salts and moisture. Gypsum-based undercoats are unsuitable, while renovation plasters are effective at dealing with salt but not moisture. Sand and cement renders are effective barriers but may not be appropriate in all conditions – especially when dealing with buildings of historic importance.

Further information

BS 6576:2005, *Code of practice for diagnosis of rising damp in walls of buildings and installation of chemical damp-proof courses.*

Property Care Association, *Code of practice for the installation of remedial damp proof courses in masonry walls,* 2008.

BRE Digest 245, *Rising damp in walls – diagnosis and treatment* (published 2007).

BRE Good Repair Guide 5, *Diagnosing the causes of dampness Year 2015.*

Subsidence

Philip Lane

Houses built on what are known as 'shrinkable clay soils' quite often have shallow foundations, usually less than 1m deep. Shrinkable clay soils are generally strong and able to support a building of four storeys on a single strip or trench-fill foundation. These soils shrink when their moisture content decreases 'subsidence' and then swell 'heave' when it increases. Slight movement of houses on foundations is therefore inevitable as a result of seasonal changes in moisture content, resulting in downward movement or subsidence occurring during the summer and upward movement or heave during the winter.

Greater movements may occur during long periods of dry weather and may lead to sticking of doors and windows and cracking of external walls. Severe movements are almost always associated with localised subsidence caused by trees whose roots extract moisture from the soil. Conversely, removing a tree tends to cause heave as moisture gradually returns to the soil. Large broad-leaved trees of high water demand are notorious for causing damage.

Before World War II, it was common practice to use shallow foundations no more than 0.45m deep. Houses built within the past 25 years should comply with guidelines issued by the National House Building Council (NHBC) and the British Standards Institution (BSI). The former requires foundation depths often well in excess of 1m, while the latter requires a minimum depth of 0.9m for any buildings founded on clay and deeper foundations where there are trees nearby.

Because of the link between clay shrinkage and the weather, insurance claims for subsidence damage increase in long dry periods. Analysis of insurance claims indicates that of those cases involving foundation movements caused by the shrinkage or expansion of clay soils, 75–80% are exacerbated by moisture abstraction by trees.

Leaking drains can also be associated with causing subsidence as a result of water washing away the soil around the foundations. This can also occur where there is an underground watercourse or spring within close proximity to the foundations.

Subsidence may also be a consequence of mining activity. The extent of subsidence due to mineral extraction depends on the method used for winning the minerals from the ground, whether by mining, pumping, or dredging. The main problems in the UK arise from coal mine workings.

In many coal fields in the UK, the presence of old workings remain as a constantly recurring problem in foundation design where new structures are to be built over them. If the depth of cover of soil and rock overburden is large, the additional load of the building structure is relatively insignificant and the risk of subsidence due to the new loading is negligible. If, however, the overburden is thin, and especially if it consists of weak crumbly material, there is a risk that the additional load imposed by the new structure will lead to local subsidence.

The risk of subsidence associated with coal workings may be obtained via a Coal Mining Report obtained from the Coal Authority.

It is important to differentiate between subsidence and settlement. In terms of foundation movement:

- subsidence is downward foundation movement caused by change in the site below the foundations, usually associated with volumetric changes of the subsoil; and
- settlement is downward foundation movement caused by an application of load usually occurring for a period of time immediately after construction or poorly compacted made ground.

Tree identification
Trevor Rushton

The influence of trees on buildings

Broadly, there are two possible mechanisms by which trees can influence low-rise buildings:

- **direct action**: the physical disturbance of the structure by root growth; and
- **indirect action**: usually associated with the changes of moisture content in shrinkable clay subsoils.

Damage due to direct action is fairly rare in buildings, although disturbance of boundary walls, brick planters, and the like is more frequent. The damage is usually related to the growth of the main trunk and roots, and will diminish fairly rapidly with distance. The radial and longitudinal pressures exerted by roots are fairly weak, and so roots will tend to grow around an obstruction rather than displace it.

Damage due to indirect action depends on a number of different factors such as the species of tree, the type of clay, proximity to the building, depth of foundations, availability of water supplies, climate, etc. The problem of clay heave or damage due to desiccation of clay soils can occur throughout the UK, although areas to the southeast of a line joining Hull to Exeter appear to be worse affected, as clay soils are very common there.

BRE Digest 298:1999 identifies four types of movement associated with clay soils and vegetation growth:

- normal seasonal movements such as would occur with a grassy area;
- enhanced seasonal movements resulting from increased moisture transpiration following the planting of trees;
- long-term subsidence occurring as a result of water deficit as trees develop; and
- long-term heave following the removal of trees and the dissipation of the water deficit.

Certain species of trees – notably oaks, poplar, elm, and willow – have been known to cause damage to buildings. A very rough rule of thumb would be to limit the proximity of a tree to its maximum potential height at maturity (the 1H rule); in many cases this will result in an overestimation of the 'safe distance' from a building.

Using a safe distance for planning purposes is one thing, but deciding whether a tree has influenced a building (particularly if the tree and the building are in different ownership) requires better data. A detailed account of the various studies that have been conducted may be found within *Tree roots in the built environment*, Research for Amenity Trees No.8, published by DCLG, 2006.

Tree identification

The accurate identification of species is usually made by reference to the characteristics of features such as the petals and stamens of flowers, as these tend to be more constant than the shape of a tree, although some basic information on the shape of the leaves can give a good indication.

Since the shape and size of a tree can be influenced by numerous factors, its shape is not necessarily a reliable guide to identification. More useful information can be provided by an examination of the bark, the twigs, and buds.

If the identification cannot be made on site, record the approximate height of the specimen and the characteristics of its bark, fruit, and leaves. Look out not only for the shape of the leaf but also its texture, its margin (see below), and its arrangement on the stalk (petiole). Are the leaves arranged in groups on a single petiole, in pairs, or singly?

Some definitions to assist identification:

- **Lamina**: the flat surface of the leaf
- **Adaxial surface**: the top surface of the leaf
- **Abaxial surface**: the under surface of the leaf
- **Petiole**: the stalk which joins the leaf to the twig
- **Margin**: the edge of the leaf – usually entire (smooth), ribbed, or lobed
- **Palmate compound leaf**: several leaflets connected to a common petiole rather like the palm of a hand
- **Pinnate compound leaf**: several leaflets, often in pairs, connected to a common petiole
- **Lenticels**: the small areas on a stem where gas exchange takes place – often small slits or round dots
- **Axil**: the angle between the petioles of adjoining leaves

As noted earlier in this section, the water demand of various species is difficult to predict with certainty, particularly due to conflicting data. The list below is in the general order of water demand but various studies show differing results and so the ranking should be treated as a rough guide only. All dimensions are approximate.

Table 4.12 Tree identification table

	**Oak** Max. height 23m. Radius of potential damage 13m. Strong root activity. Deeply fissured bark. Lobed, simple leaves, 5–7 lobes per side. Buds clustered near end of twig. Drooping catkins on male trees; acorns on female.
	**Poplar** Max. height 24m. Radius of potential damage 15m. Strong root activity. Simple palmate leaf with downy abaxial surface.
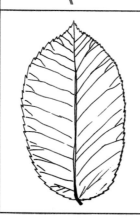	**Elm** Max. height 25m. Radius of potential damage 12m. Strong root activity. Variable size simple leaf, some with shiny leaf and hairs on abaxial surface.

Ash
Max. height 23m. Radius of potential damage 10m. Medium root activity, fast growing – up to 500mm per year. Compound, pinnate leaf. Pale grey bark with ridges.

False acacia
Max. height 20m. Radius of potential damage 8.5m. Medium root activity. 13 leaflets. Dull brown bark, often deeply furrowed. Two spines at each bud.

Horse chestnut
Max. height 25m. Radius of potential damage 10m. Medium root activity. Compound palmate leaf. Bears fruit in late summer/autumn.

Hawthorn
Max. height 10m. Radius of potential damage 7m. Simple leaves with deep lobes. Thorns on twigs. White blossom in spring.

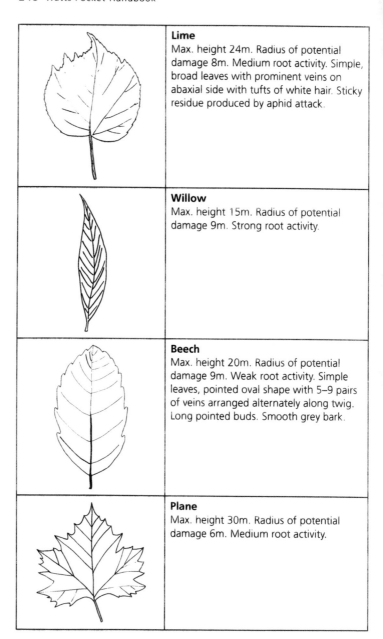

Lime
Max. height 24m. Radius of potential damage 8m. Medium root activity. Simple, broad leaves with prominent veins on abaxial side with tufts of white hair. Sticky residue produced by aphid attack.

Willow
Max. height 15m. Radius of potential damage 9m. Strong root activity.

Beech
Max. height 20m. Radius of potential damage 9m. Weak root activity. Simple leaves, pointed oval shape with 5–9 pairs of veins arranged alternately along twig. Long pointed buds. Smooth grey bark.

Plane
Max. height 30m. Radius of potential damage 6m. Medium root activity.

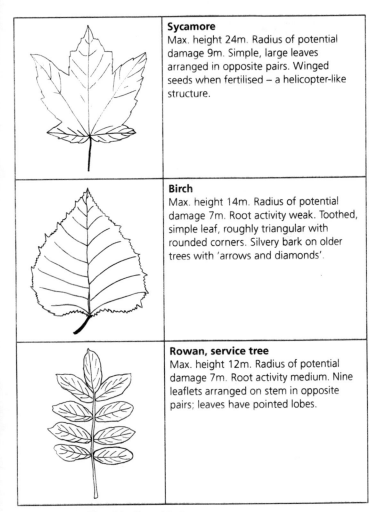

	Sycamore Max. height 24m. Radius of potential damage 9m. Simple, large leaves arranged in opposite pairs. Winged seeds when fertilised – a helicopter-like structure.
	Birch Max. height 14m. Radius of potential damage 7m. Root activity weak. Toothed, simple leaf, roughly triangular with rounded corners. Silvery bark on older trees with 'arrows and diamonds'.
	Rowan, service tree Max. height 12m. Radius of potential damage 7m. Root activity medium. Nine leaflets arranged on stem in opposite pairs; leaves have pointed lobes.

Drawings contributed by Fiona Mackay

Further information

There are several useful online guides to assist identification, for example:

www.woodlands.co.uk/owning-a-wood/tree-identification/
http://apps.kew.org/trees/?page_id=17

4.3 Cladding

Mechanisms of water entry
Trevor Rushton

Solving problems of moisture entry into a building can be troublesome, as the sources of the problem can be difficult to identify and the paths taken obscure. Following the pathway from suspected entry to exit is a useful technique; understanding the essential mechanisms of water entry and transfer can help diagnosis.

Mechanisms of water entry

Rainwater can penetrate a building in the following ways:

- kinetic energy;
- surface tension;
- gravity;
- capillarity;
- pressure differentials; and
- any combination of these.

Kinetic energy

This is the direct action of the wind carrying a droplet of rainwater with sufficient momentum to force it through a sealed joint. Prevention or drainage overcomes this. Prevention can be by baffles, by a durable seal, or by a labyrinthine shape within the joint. Drainage collects the penetrating water and diverts it back to the outside.

Surface tension

This can cause water to adhere to and move across surfaces. It is guarded against by drip edges or throatings along leading edges, and horizontal surfaces should slope down and out. Connecting components can also have appropriate grooves or ridges.

Gravity

This can take water through open joints that lead inwards and downwards. Reversing the slope overcomes this.

Capillarity

This occurs in fine joints between wettable surfaces. It is only severe when other mechanisms persist, for example, wind-assisted capillarity. In metal components it is resolved by capillary breaks within the joint surfaces.

Pressure differentials

This is a very important mechanism. Since air at high pressure will migrate to areas of low pressure, it follows that high pressure air can transport moisture into an area of lower pressure. It is overcome by maximising the outer deterrent and minimising the pressure differentials. This is achieved by self-contained (compartmentalised) air spaces behind the outer skin, which are well ventilated to the outside thus enabling rapid equalisation of pressures.

Variations in vapour pressure work in the same way. Vapour pressure is the pressure exerted by water vapour (a gas) being just one of the components of air. Air with a high vapour pressure will naturally move towards and area of lower vapour pressure, the rate at which it can achieve equilibrium depends upon the vapour permeability of the building materials through which it must first pass. Materials such as metal foil are highly vapour impermeable while plasterboard for example is far more permeable.

Curtain walling systems
Trevor Rushton

'Curtain walling' is a weatherproof and self-supporting enclosure of windows and spandrel panels in a light metal framework which is suspended right across the face of a building, held back to the structure at widely spaced joints.

Types of curtain walling systems

Stick systems

Stick construction is the traditional form of curtain walling, comprising a grid of mullions and transoms into which various types of glass and/or insulated panels can be fitted. Most of the grid assembly work is done on site. The advantages include relatively low cost and the ability to provide some dimensional adjustment. The disadvantage is that performance is workmanship sensitive. It is not unusual to find systems failing initial waterproofing tests during erection.

Unitised systems

Unitised systems comprise narrow-width storey-height units of aluminium framework containing glazed and/or opaque panels. The entire system is pre-assembled under factory-controlled conditions. Mechanical handling is required to position, align, and fix units on site onto pre-positioned brackets attached to the floor slab or the structural frame. Modern installation techniques increase the speed of erection and often minimise the requirement for scaffolding. Unitised systems have higher direct costs and are less common than stick systems. Nowadays the curtain walling to most prestige buildings is of this type.

Panellised systems

Panellised curtain walling comprises large prefabricated panels of bay width and storey height which are connected back to the primary structural columns or to the floor slabs. Panels may be of precast concrete or comprise a structural steel framework which can be used to support a variety of stone, metal, and masonry cladding materials. The advantages of these systems are improved workmanship as a consequence of factory prefabrication, allowing improved control of quality and rapid installation with the minimum number of site sealed joints. Panellised systems are less common and more expensive than unitised construction; however, they often appear similar to unitised systems.

Variations

Structural sealant glazing: this is a form of glazing that can be applied to stick or unitised curtain walling systems. With structural sealant glazing, the double-glazed units are attached to the grid framework preferably with factory-applied structural silicone sealant rather than by pressure plates and gaskets in a more traditional system. The attraction of this form of glazing is that it provides relatively smooth facades which are visually attractive. Some systems involved bonding the glass to a metal carrier system that is itself fixed mechanically to the supporting framework. It is usual for structural sealant glazing to be bonded on two sides and mechanically fixed on the top and bottom edges – 'two-sided' although some systems involve 'four-sided' fixing.

Structural glazing: this typically comprises large, thick, single panes of toughened glass assembled with special bolts and brackets that are supported by a secondary steel structure. This form of glazing is often referred to as 'planar' glazing and is commonly used to form the enclosure to atriums and entrances.

Weather tightness

There are various methods of preventing rainwater ingress:

Face-sealed systems

Early curtain walling systems tended to be face sealed, relying on a weatherproof outer seal to prevent water penetration. The seal must remain completely free of defects to prevent leakage paths occurring. Where there is no provision for drainage, any water that bypasses the outer seal could result in internal water ingress. Given the difficulty in ensuring 100 per cent weather tightness, face-sealed systems are rare; most systems now aim to tolerate a small amount of leakage and to ensure that it can be discharged harmlessly without penetrating the interior of the premises.

There are also a small number of proprietary systems incorporating front zipper gaskets containing a large rubber gasket with a central press-in segment or zip which, when pressed into place, forces the gasket out onto the surface of the glass. This system does not normally have provision for water drainage.

Fully bedded systems

These systems are now largely obsolete and were used on early forms of curtain walling. Fully bedded glazing is a face sealed system relying on the glazing rebate being completely filled with glazing compound to prevent the passage of water. They are therefore 'undrained'. Any voids within the bedding are a potential weak link for water ingress and early failure of the double-glazed units.

Drained and ventilated systems

Most cladding designers now accept that it is difficult to exclude water and therefore provision for a small amount of leakage can be made within a drained system. Typically these dry-glazed systems comprise an outer decorative cover plate, and an aluminium pressure plate with two narrow rubber oyster gaskets on either side clamped against the glass or insulated infill panel. The pressure plates are screw-fixed through a thermal break into the mullion or transom member. A further inner gasket between the glass and the mullion or transom provides a further seal.

In drained and ventilated systems, the front gaskets provide an initial barrier. The rebates and cavities are drained and ventilated to the exterior to prevent the accumulation of any water that bypasses the outer seals. Drainage is usually via small holes or slots in the underside of transoms that drain water down through the mullions.

Some systems also incorporate a foil-faced butyl adhesive tape applied over the transom and mullion nosings directly beneath the pressure plate to serve as a secondary line of defence.

While these systems will accommodate a small quantity of water within the glazing rebates, it is important that the pressure plate is fixed to the correct torque so that the gasket seals form a good seal against the glass.

While structurally bonded systems appear to rely on silicone seals between the glass, they will often contain drainage and ventilation systems in the same way as conventional curtain walls.

Pressure-equalised systems

Pressure-equalised systems are an improved variation of drained and ventilated systems. Here the ventilation openings in the pressure plates are of an increased size to permit rapid equalisation of pressure in the glazing rebates with the external pressure thereby preventing water penetration of the outer face. Consider the following diagram.

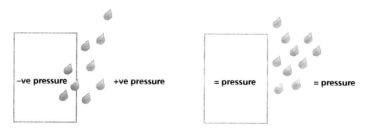

Figure 4.4

The rectangular box may be considered as the area around the glazing in a cladding system. If the pressure inside the box is less than the pressure outside, the water will be drawn in. If we can equalise the pressure in the box to that outside the box, then the probability is that water will stay on the outside. In curtain walling systems we must provide a perfect seal around the inside of the window and must also provide the number of slots around the perimeter of the glass to enable pressure within the glazing rebate to equal that of the external air pressures almost instantaneously.

With pressure-equalised systems, the outer face is sealed as tight as possible against rainwater while the inner face is sealed as tight as possible against air inflow. Nearly all modern curtain walling systems are designed utilising the principles of pressure equalisation.

It is very difficult to identify the differences between a drained and ventilated system and a pressure-equalised system. However, for pressure equalisation to work properly, the various zones of pressure must not be too large. Thus it is common to consider the area around one glass pane as one zone, and therefore drainage must be made from the transom members and not from the mullions. In practice it is very difficult to provide full pressure equalisation, and there is some doubt in the industry as to whether it is fully effective.

Double-glazed units (insulating glass units) (DGUs and IGUs)

Double-glazed units comprise two or more panes of glass spaced apart and hermetically factory-sealed with dry air or other gases in the unit cavity. The perimeter edge sealant prevents moisture from entering the unit cavity and holds the unit together. Two forms of edge seal configuration are single-seal systems and dual-seal systems. Single-seal systems rely on the edge sealant to act both as the vapour barrier and as an adhesive bond to hold the panes of glass together. Dual-seal glazing units rely on two seals, an inner seal to control water vapour transmission and an outer secondary seal to hold the glass tightly against the spacer bar. The combined properties of primary and secondary sealant provide high-quality glass units. Nearly all units are now of this type.

Types of glass

The following types of glass are used in curtain walling systems:

Annealed glass: this is untreated glass manufactured from soda lime silicates. It is the least expensive and most readily available type of glass. Annealed glass breaks into sharp edged shards and is therefore considered to be unsafe in all fire and breakage situations.
Thermally toughened glass: this is formed by heating and then rapidly cooling or quenching annealed glass. Differential cooling and hardening across the thickness of the glass generates a compressive stress in the surface layer of the glass. Toughened glass is always a safety glass and compared with annealed glass is four to five times stronger in compression and bending. In failure,

toughened glass shatters into small, relatively safe fragments. Toughened glass can suffer from a problem known
as 'roller wave' – this produces optical distortions that can
be particularly noticeable in mirrored glass claddings.

Heat-strengthened glass: this is formed by heating
annealed glass and then cooling it under controlled conditions. Heat-strengthened glass offers some of the strength
of toughened glass but a reduced risk of failure due to
nickel sulphide inclusions because of the reduced tensile
stress in the glass. This glass is often also referred to as
'partially toughened'. Heat-strengthened glass is also less
likely to display roller wave distortions.

Laminated glass: this is formed by bonding together two
or more panes of glass using a plastic (polyisobutylene
[PIB]) interlayer. Any of the above forms of glass may be
used in any combination. Upon failure, laminated annealed
glass breaks into shards which are held together by the
interlayer. Laminated glass may include one or more panes
of toughened glass. If all panes are of toughened glass
then the broken glazing will lose all structural integrity
and may pull free from the pane unless properly secured.
Laminated glass is recommended for the inner pane of
overhead glazing and is considered to be safety glazing.

Low-e coatings: these reduce the emission of long-wave
thermal radiation from the glazing and increase the reflection of this radiation. In the winter, solar radiation can be
trapped within a room and reduce the need for heating. In
the summer, however, heating can also occur and so Low-e
glass needs to be used in conjunction with adequate provision for ventilation. Most Low-e glass can be toughened
and laminated. The coatings can either be 'soft' or 'hard'
(pyrolytic) coatings – the latter generally being more robust.

Surface finishes

The metal components of a curtain walling system will nearly
always require finishes to provide protection against corrosion
or for appearance.

To preserve the decorative and protective properties of any metal finishing, it is essential that atmospheric deposits are removed at frequent intervals, particularly those surfaces which are not exposed to the washing effects of the rain. If the finish becomes chalky, specialist cleaning systems can be used, but these are very expensive.

Beware of spray applied repairs to site damage during installation; these areas will weather at a different rate and become visually apparent within a few years.

Quality of workmanship is particularly important and it is therefore essential to choose a reputable applicator preferably covered by a quality insurance scheme. Independent acceptance inspection testing can be undertaken to ensure compliance with the specification.

Organic coatings (carbon-based coatings)

Organic coatings are normally applied to either steel or aluminium and include polyester powder, PVDF, PVC, Plastisol, and polyester. The most common organic finish for windows and curtain walling is polyester powder coating; however, a range of wet-applied finishes is widely used for opaque cladding panels. Polyester powder coatings may be applied to either galvanised steel or aluminium and are available in a wide range of colours. They are tough and abrasion resistant. Manufacturers often provide a guarantee for 15–20 years.

Anodising

Anodising is an electrolytic process that produces a dense, hard, and durable oxide layer on the surface of aluminium. The oxide layer is porous and must be sealed to prevent staining, but can be coloured by introducing dyes or chemical treatment before sealing. Anodised finishes are generally harder and more abrasion-resistant than organic coatings, with an expected life of 50 years or more. However, anodised surfaces are susceptible to alkaline corrosion from contact with fresh concrete or mortar and rainwater run-off from concrete surfaces.

Testing

Many standard and bespoke curtain walling systems are tested in laboratory conditions to determine the resistance to wind load, airtightness, and water tightness. These tests are undertaken on a very small number of test panels assembled in factory conditions. By necessity they are a test of the design rather than on-site workmanship.

It is therefore critical that all installed curtain walling systems are subject to on-site testing to establish that the fabrication and installation has been undertaken to a satisfactory standard.

Water tightness on site is typically assessed using the methods outlined in the following documents:

Pressure spray in accordance with AAMA standard 501-94, *Methods of test for exterior walls*, Architectural Aluminium Manufacturers Association, USA
CWCT, *Test methods for curtain walling*, 1996
BS EN 13051:2001, *Curtain walling. Watertightness – Field test without air pressure using a water spray bar*
BS 5368-2: 1980, *Methods of testing windows. Watertightness test under static pressure*
ASTM E1105-96, *Standard test method for field determination of water penetration of installed exterior windows, curtain walls and doors by uniform and cyclic static air pressure difference.*
The CWCT and AAMA test regimes are essentially the same

Ideally, the first areas to be tested should be among the first areas of each type of curtain wall to be constructed on site. Typically, the test areas are at least one structural bay wide and one storey in height, providing that all horizontal and structural joints or other conditions where leakage could occur are included.

Under the CWCT and AAMA methods, water is applied via a brass nozzle on the end of a hose that produces a solid cone of water droplets with a spread of 88°. The nozzle is provided with a control valve and a pressure gauge between the valve and nozzle. The water flow to the nozzle is adjusted to produce 22 +/– 2 litres per minute, producing a water pressure at the nozzle of 220 +/– 200 Kpa (Kpa = 1000 pascals).

Water is directed at the joint perpendicular to the face of the wall and moved slowly back and forth over the joint at a distance of 0.3m from it for a period of five minutes for each 1.5m of joint. There should be an observer of the inside of the wall, using a torch if necessary, to check for any leakage.

Further information
BS EN 13830:2015 Curtain walling – Product standard
BS EN 12152:2002 Curtain walling – Air permeability – Performance requirements and classification
BS EN 14019:2004 Curtain walling – Impact resistance – Performance requirements
www.cwct.co.uk/publications/list.htm The Centre for Window and Cladding Technology is the UK's premier authority on cladding and curtain walling. CWCT publishes extensive guidance although this is generally only available to purchase.

Spontaneous glass fracturing due to nickel sulphide inclusions

Trevor Rushton

Nickel sulphide is one of several chemical contaminants that can occur during the manufacture of glass. There is some debate as to its origin, but it is thought that it is due to the mix of nickel and sulphate impurities within the glass batch materials, the fuels or even the furnace equipment, and this creates polycrystalline spheres which vary from microscopic to 2mm in diameter.

All glass has some of these inclusions present; they are impossible to eliminate entirely and therefore they are not considered a product defect.

In untreated (annealed) glass they are not a problem. But when glass is heat treated (toughened or tempered), the inclusions are modified into a metastable state. The particles initially decrease in volume, but over a period of time (sometimes years) gradually revert to their original volume; this expansive force acts on the glass around the particle.

In a majority of cases this has little effect but, dependent on size and proximity to the centre of the pane where the forces are greatest, this can eventually cause the glass to break. Failures can occur with inclusion sizes in the 100–200 micron range, but sizes above this are far more likely to cause failures.

There is a theory that for an initial period of approximately 1 year after manufacture there are relatively few breakages. After this, the number increases for up to several years, thereafter decreasing in frequency. There have been reported incidences where fractures have occurred more than 20 years after the installation of glass.

Panes in external situations are at greater risk. However, there have been a small number of cases where spontaneous breakage has occurred in internal glazing, remote from external influences, for example, panels to a staircase balustrade or an internal partition.

Because of the risk of falling debris some toughened glass should be avoided in sloped overhead applications where it is used either as a single pane or as the inner leaf of a sealed unit. If laminated toughened glass is used, the polyisobutylene (PIB) interlayer and the intact sheet may be more likely to hold the glazing in place.

In heat-strengthened glass, nickel sulphide inclusion is not generally regarded as a source of fracture. The difference between this and toughened glass is the rate of cooling. In the former this is less rapid, reducing surface compressive strength and making it much less susceptible to the transformation of nickel sulphide inclusions. Offering 5.5 times the strength of annealed glass, in many circumstances it is a useful replacement for tempered glass. However, it is not a suitable substitute where safety glass is required.

'Heat soaking' is a quality controlled process which gives increased reassurance against the presence of critical nickel sulphide inclusions by subjecting the glass panels to accelerated elevated temperatures to stimulate the transformation of the crystals and thus initiate immediate failure. It is thought that this process identifies 90% or more of the glass which might have subsequently failed after installation. The heat soaking process could be used either as a sampling method or as an additional treatment, which in the case of clear toughened glass could add up to 20% to the cost.

Heat soaking does not change any of the physical properties of toughened glass, and therefore there is no means of distinguishing whether or not this process has been carried out. Current best practices dictate that specifiers should ensure that they specify 'heat soaked toughened glass to pr EN 14179'. This standard reduces the anticipated failure rate to 1 per 400 tonnes of glass used. Prior to EN 14179, heat-soaking standards were less rigorous, leading to unreliable results; a breakage rate of 1 per 5 tonnes of glass was thought an average level. However, this corresponds to only 200m^2 of 10mm glass, and so a glazed roof of 6,000m^2 could be expected to have 30 breakages.

It is generally accepted that nickel sulphide contamination is a problem that can affect batches of glass – some batches may perform perfectly well, while others suffer a high proportion of breakages. Examination of the edge spacers of double-glazed units may reveal information as to the age of the glass panel (usually month and year of manufacture) although it may not contain information as to the actual production run of the glass.

Identification

When toughened glass is broken, the tensile stress is spread out from the source causing the pane to crack into small fragments (dicing). These fragments tend to be slightly wedge-shaped, emanating from the source of the fracture and are often held into position wedged against the frame due to their increased volume.

If the fracture is as a result of expansion of the nickel sulphide inclusion, those fragments immediately adjacent are more hexagonal and at the epicentre of the breakage the two larger particles form a distinctive butterfly shape linked by a central straight line crack. If large enough, the inclusion may be seen in the form of a black spec, or its presence may be confirmed by optical or SEM microscopy.

When carrying out an investigation, all possible causes of failure should be considered, including poor glazing toler-ances and insufficient allowance for subsequent movement of the frame and any supporting structures. Possible causes may include deflection or rusting of steel frame, shrinkage of concrete frame, thermal movement, normal air pressures, and even sonic booms. If the fracture is a result of impact or of local point loading, there should be evidence of local crushing.

The chances of installing a toughened glass pane which may later fail due to the expansion of nickel sulphide inclusions are very small. Where it is essential, for reasons of accessibility or safety, that the pane should not fail, alternative forms of glass should be considered. One particular example would be in overhead situations – say in a shopping mall.

Testing

Visual testing of undamaged glass is not practicable; photographic examination has been shown to be possible but is not commercially viable.

Samples of fractured glass should be sent intact with the epicentre protected with clear film to a suitable testing laboratory. Microscopic analysis will be sufficient for an initial diagnosis, but a scanning electron microscope will yield more information.

Further information

BS EN 14179-2: 2005 *Glass in building. Heat-soaked thermally toughened soda lime silicate safety glass. Evaluation of conformity/product standard.*
Report 471, *Sloping glazing: understanding the risks*, BRE, 2004.
Special Publication SP 161, *Guidance on glazing at height: An introduction for the client*, CIRIA, 2005.

Composite panels
Trevor Rushton

After a series of major insurance losses in the late 1990s and the early part of the C21, insurers sought to limit their risks by insisting that composite panels, and particularly those with thermoplastic cores, be replaced with materials of known fire performance. Generally, materials containing polyurethane (PUR) cores were considered unsatisfactory, while polyisocyanurate (PIR) cores that complied with Loss Prevention and Factory Mutual standards were favoured. As an alternative to replacement, insurers charged inflated premiums or increased the level of deductibles. Even though materials complied with Building Regulations, the insurance conditions could be prohibitively expensive.

More recently, the insurance industry has been less stringent, but for surveyors it is often best to attempt to identify the core so that appropriate advice can be given. The existence of composite panels needs to be kept in perspective; the main risks arise when the building is constructed with a large proportion of composite panels; isolated sections or small areas of panel in proportion to the whole are unlikely to make a huge difference to the fire load.

Composite, or sandwich, panels comprise two outer layers of steel or aluminium sheet bonded to an inner core of insulating material. The resulting product is lightweight yet strong and able to span greater distances than would be possible with the individual component parts in isolation. Flexural strength is attained by maintaining the bond between the layers. One side will be in tension, the other in compression. Remove the bond to one face and integrity is destroyed and the panel will fail.

Typical insulating cores and their properties are given below:

Table 4.13

Insulating Core	Properties
Foamglass (fairly rare)	Non-combustible
Thermosetting rigid foam – polyisocyanurate (PIR) and polyurethane (PUR)	Undergo localised charring, although flaming can take place if flammable vapours are released. The charred material will shrink and can lead to delamination of panels. While PIR is also a thermosetting foam, it performs better in fire than straight PUR. Many pre-2000 panels were of PUR while today PIR is the material of choice.

Insulating Core	Properties
Thermoplastic polystyrene (EPS);	Polystyrene materials may burn fiercely, give off thick black toxic smoke and allow burning droplets to fall.
Machine made mineral fibre	Generally non-combustible, but some of the adhesives used to bond the panels can be combustible.
Phenolic foam	Generally non-combustible, but some types may hold water and have a corrosive effect upon the metal outer skins.

Panel instability in a fire can affect composite panels that are not properly supported or restrained. In the case of roof and wall cladding, there are usually additional supports in the form of purlins and sheeting rails and primary fasteners, which serve to tie the two leaves together and prevent them from becoming detached. However, in internal situations (such as food processing plants, cold storage facilities, etc.) the quantity of insulation required will often lead to panels of 200mm thickness or more. Such panels require less support and so present a greater risk of delamination. It is these panels which cause the greatest levels of concern – particularly as the core materials are often EPS.

There will be a conflict between the requirements of the Building Regulations and the requirements of insurers. Building Regulations are aimed essentially at ensuring the health of users and neighbours of the building (and of course visitors or the emergency services). Insurers may want higher standards of protection, fire suppression and/or more reliance upon non-combustible materials.

The relevant standards for composite panels were originally set by the Loss Prevention Council but are now administered by BRE Global under LPS 1181. Generally, materials with a PUR or EPS core will almost certainly not satisfy the requirements of LPS 1181, whereas PIR or MMMF products can be engineered to comply. Note that compliance is not simply a case of the correct core material, edge jointing details are also critically important.

The Association of British Insurers has produced a report on the issue and it is hoped that insurers will now take a more relaxed view if buildings have been constructed using materials certified under LPS 1181 Part 1 (for external systems) or Part 2 (for internal applications). However, many buildings have been constructed (and still are constructed) using materials that do not satisfy these standards and, in these circumstances, it is important to consider the overall level of risk rather than the mere existence of the panels.

Whether or not sandwich panels constitute a risk is a matter of judgment and scientific fire risk assessment. A reasoned approach may involve the consideration of the following:

Is there a sprinkler installation in the building?
Are there any specific fire risks – use, storage of inflammable materials, arson, etc.?
Are the panels in the vicinity of battery charging areas?
Are the panels perforated such that the cores are exposed?
How are the panels fixed – are they properly restrained?
What is the nature of the insulant?
What is the extent of the material and to what extent could it contribute to fire load?

Many manufacturers keep records of consignments and can often, given the nature of the contractor, identify the nature of the material supplied to a specific site. More modern trends have included a small identifier that can be marked on the panels and revealed by exposure to a small UV light source.

Further information
BS 8414-2:2015: *Fire performance of external cladding systems. Test method for non-loadbearing external*

cladding systems fixed to and supported by a structural steel frame (incorporating corrigendum No. 1). Series/doc. No Year 2015

Loss Prevention Standard 1181 Part 1: Issue 1.2, *Series of fire growth tests for LPCB approval and listing of construction product systems. Part 1: Requirements and tests for built up cladding and sandwich panel systems for use as the external envelope of buildings,* Issue 1.2. 2014.

5
Sustainability

5.1 Environmental

Contaminated land
Janette Stevens

Environmental due diligence is now commonplace in property transactions. The presence of contaminated land can adversely affect site value and rental income and can hinder transactions if not properly managed. With an increasing move away from greenfield development, it is no longer possible for the majority of investors or tenants to avoid owning or occupying some land affected by contamination, whether as a city-centre property that has had a variety of past uses or a new out-of-town development constructed on an old industrial site.

Statutory definition of contaminated land

Under Part IIA of the Environmental Protection Act 1990 'contaminated land' is defined as:

> *'land which appears to the Local Authority in whose area it is situated to be in such a condition, by reason of substances in, on or under the land, that:*
>
> a. significant harm is being caused or there is the significant possibility of such harm being caused; or

b. significant pollution of controlled waters is being caused, or there is a significant possibility of such pollution being caused'.

©Crown copyright material is reproduced under the Open Government Licence v1.0 for public sector information: www.nationalarchives.gov.uk/doc/ open-government-licence/

This definition was amended on 6 April 2012 through the Water Act 2003 (Commencement No. 11) Order 2012. This was in response to widespread criticism of the original guidance. Under the original guidance, councils commonly had difficulty assessing whether land was sufficiently contaminated in order for action to be taken to enforce clean-up. The knock-on effect was that redevelopment was often delayed with significant financial implications. The Water Act 2003 therefore amended the definition of contaminated land to refer to the 'significant' pollution of controlled waters in order to address the uncertainty.

Many sites that are contaminated will not fall within the definition and will not be classified as 'contaminated land' under Part IIA. However, such contamination could still have implications for owners and occupiers (e.g. in terms of liability affecting the saleability and marketability of a site) and may still require a Phase II investigation.

Land is only defined as 'contaminated land' if there is a 'significant pollutant linkage' present. There must be evidence of a 'source – pathway – target' relationship. This means there should be a source present, a receptor that could be harmed by the source (e.g. humans), and a pathway linking the two.

Documentation

Documentation is the key to ensuring a smooth property transaction. Documentation should be in place to demonstrate that the site condition has been adequately assessed. This can be undertaken by commissioning an Environmental Audit and/or an intrusive ground/groundwater investigation.

These assessments, which are discussed in further detail below, should ascertain:

- whether the site condition has been adequately assessed by an environmental assessment or review (commonly known as either an Environmental Audit or Phase I) and/ or an intrusive site investigation (Phase II) and there are no significant information gaps;
- whether contamination is present (or is likely to be present), and the types of contaminants;
- where contamination has been identified, it does not represent a risk to the existing or proposed use of the site;
- that contamination is not migrating off site within groundwater;
- that contamination does not represent a significant risk to groundwater and surface water resources or other sensitive receptors (e.g. sites of special scientific interest);
- that contamination does not represent a risk of regulatory authority action (e.g. under Part IIA of the Environmental Protection Act 1990);
- that contamination does not represent a risk of third party action (e.g. from adjoining land owners); and
- whether any potential liabilities may exist from any forthcoming business transaction.

Environmental Audit

While an Environmental Audit desktop study does not include a site visit, an Environmental Audit (also known as a Phase I or an Environmental Assessment) is based on background research and should include a site inspection or walkover. A walkover would normally be advised to ensure all present day site issues are appropriately assessed. The research will normally include a review of:

- current site uses;
- historical site activities;
- environmental sensitivity;
- regulatory authority records; and
- a risk assessment and environmental risk rating.

Intrusive ground/groundwater investigation

An intrusive ground/groundwater investigation (also known as a Phase II) is based on a physical assessment of the underlying site conditions, and usually comprises chemical analysis and/or monitoring of soil, groundwater, and ground gas.

If an Environmental Audit or existing knowledge identifies a potentially significant contamination issue, then it may be necessary to conduct a Phase II investigation to gain an understanding of whether contamination is actually present and whether it is likely to represent a significant risk or liability.

A Phase II investigation will normally include:

- a summary of the Environmental Audit findings;
- a description of the Phase II investigation methods;
- data and observations recorded during the site work, including field evidence of contamination;
- data from chemical analysis of ground/groundwater/ ground gas samples;
- interpretation of the results and an assessment of risk; and
- recommendations for remediation (if required) of the underlying ground and/or groundwater.

The Phase II investigation should be designed to address the specific issues raised by the Environmental Audit, such as the range of contaminants highlighted as possibly being present (e.g. hydrocarbons at a petrol station, ground gas from a landfill site, etc.).

Typical sampling methodologies are:

- **Soil sampling**: Soil samples are required for chemical analysis to determine the presence of any contamination. Methods of soil sampling differ depending on the contaminants being tested. Usually disturbed samples are adequate for testing most chemicals. These can be obtained from the excavator bucket when trial pits are being used, from the cuttings from boreholes or from the window within a window sampler. Samples are analysed in the laboratory for a wide range of 'baseline' contaminant chemicals, often supplemented by further specialist testing depending upon the type of contamination present.
- **Water sampling**: Water samples can be obtained by inserting a standpipe into a completed borehole. Water samples will be taken after insertion for further chemical analysis. The well may also be used for the monitoring of gas. Installation can be permanent or semi-permanent to facilitate further sampling at a later date. Laboratory analysis is similar to soil samples.
- **Gas sampling**: The presence of methane and carbon dioxide may be established on sites which are landfills or close to neighbouring landfills. Monitoring wells constructed for water sampling may also be used to sample the presence of land gas. In-situ measurements of gas concentrations and flow are taken at ground level on the head of the monitoring well using an infrared gas analyser. Wells may be constructed to measure gas concentrations and flow at different depths.

The majority of sites undergoing development will require a Phase II investigation, particularly where the previous use was industrial. Phase II investigations can often be combined with geotechnical investigations for foundation design. Sites that are not being developed may also require Phase II investigations (e.g. during transactions). This may depend on whether a significant pollutant linkage has been identified or if the local authority may investigate the site as part of their Part IIA inspection strategy in the near future.

Remediation

Remediation of sites can be achieved by removing sources of contamination, reducing levels of contamination, or modifying the 'pathways' between the source and the identified sensitive receptors. The main approaches include:

- excavating and removing contaminated material off site to landfill (dig and dump);
- encapsulating or separating contaminated material on site by severing contaminant pathways with barriers (e.g. underground bentonite walls); and
- treating the contaminated material, either in situ or after removal (e.g. bioremediation or soil washing).

Remediation can often be combined with the redevelopment of the site. Remediation strategies should be approved in advance by the local authority and the Environment Agency. Post-remediation validation sampling (e.g. of soil or groundwater) should be undertaken to document that the remediation has been effective.

Environmental insurance

Environmental insurance is increasingly being used in property transactions to cover risks associated with contaminated land. Environmental insurance can be obtained directly from specialist underwriters or through an insurance broker. A broker will normally obtain quotations from a number of different underwriters in order to negotiate the best insurance cover and premium for a client.

The most common type of environmental insurance covers regulatory and third party claims due to land contamination. Other types of insurance can provide protection against unpredictable costs should site remediation expenses prove difficult to quantify at the planning stage.

When taking out environmental insurance, one of the most important aspects is to understand what circumstances the policy would cover. Furthermore, as with many insurance policies, environmental insurance policies are often written in a language that is sometimes unclear.

Common exclusions from policies are:

- known contamination,
- business disruption costs,
- remediation costs on change of land use, and
- loss in value.

Key contaminated land legislation

The government believes that the planning process provides the best means of remediating sites. In most cases the planning system is capable of handling such issues, for example by requiring Phase II investigations or by requiring remediation to be approved by the local authority prior to development. A new contaminated land regime was introduced in 2000 in the UK (as Part IIA of the Environmental Protection Act 1990). This is intended to deal with problem contaminated sites that are not being developed and which would therefore not be dealt with under the planning system.

Enforcing authorities

The local authority is duty bound to provide a public register containing information about land that has formally been identified as potentially contaminated, and the action which has been taken to remediate it. Any land that has satisfactorily been remediated prior to a remediation notice being served will not appear on the public register. Furthermore, if land is to be redeveloped in the near future it is unlikely that a notice will be served.

The cost of any clean up will normally lie with the person (or 'appropriate persons') who knowingly caused the contamination. The local authority will ensure the clean up is carried out either through the planning process, via voluntary remediation, or if necessary by serving a remediation notice requiring them to clean up the site. In the case of an emergency the council will remediate the site and recover the costs afterwards.

Local authorities have primary responsibility for the identification of contaminated land, although the Environment Agency (EA) will respond to requests from a local authority for information on land that it is considering prioritising for inspection. Where a site has been determined as contaminated land, local authorities must take into account any information held by the Environment Agency on issues of water pollution.

Special sites

Land may also be categorised by a local authority as a 'special site'. This definition includes sites which may be affecting underlying water supplies or major aquifers, such as military land, oil refineries, and nuclear plants.

Once the local authority has made this designation then the Environment Agency takes over the enforcing role. The Environment Agency would then maintain full responsibility for the site including both the cost of enforcement and also any testing and monitoring required. Should the site become an 'orphan site' (when no knowing polluter is found) then the Environment Agency will also be responsible for the cost of remediation. The disadvantage from the local authorities' perspective is that they do lose control of what can be politically sensitive sites. If there is a dispute as to whether or not the site is a special site, the matter can be referred to the Secretary of State for determination.

Remediation notice

The local authority cannot serve a remediation notice until 3 months have elapsed since the person or persons were notified of the designation of their land as contaminated. If the local authority, in the course of the consultation period, finds additional appropriate persons, it must notify them of the designation of the land as contaminated and then wait a further 3 months before serving any remediation notice on them.

Although the local authority is under a duty to serve a remediation notice, it may wait for more than 3 months before doing so. This may well be the case where discussions about voluntary remediation are ongoing. It is vital that time limits be set for the actions that are going to be required otherwise it will not be possible for the local authority to initiate enforcement proceedings on the basis that the action has not been carried out.

Any remediation notice must effectively be justified with reference to the statutory Contaminated Land (England) Regulations 2006 (SI 2006/1380) guidance by including:

- the remediation scheme proposed,
- any exclusion from liability, and
- any apportionment of costs.

The remediation notice may, therefore, have to be a fairly lengthy document. It should be carefully drafted so as to attempt to avoid an appeal being made against it. Remember that there might be several remediation notices for each site. Copies should be provided to all the parties that were consulted about remediation, and to the Environment

Agency. Where there are several appropriate persons for a given action, a single remediation notice may be served on all of them. Details of the notice must be included in the register.

Appeals

There is a right of appeal against a remediation notice. Where the local authority serves a remediation notice the appeal is heard in the local magistrates' court. An appeal is made by way of a summary application to the court. If the Environment Agency serves the remediation notice because it has taken over regulation of a site and it is now the appropriate authority, an appeal will be heard by an inspector appointed by the Secretary of State.

The time limit for bringing an appeal is 21 days, beginning with the first day of service. A remediation notice, on appeal, can be modified, confirmed, or quashed. The remediation notice may be quashed if there is 'a material defect' in the notice. The grounds for appeal are set out in the Contaminated Land (England) Regulations 2006. There are a large number of grounds of appeal including:

- the appellant is not the appropriate person,
- the authority failed to exclude the appellant,
- there has been an improper apportionment of costs,
- there is some error with the notice,
- the requirements of the notice are unreasonable having regard to the costs and benefits, and
- the period of time for compliance is insufficient.

Who is liable under Part IIA?

Remediation notices are served in the first instance on 'Class A' persons, i.e. polluters or knowing permitters of contamination. If these responsible parties cannot be found, the current owner or occupier may be responsible (although for a more limited range of liabilities) – 'Class B' persons. Several parties may be implicated.

Sellers can avoid liability where there are payments for remediation, with an explicit statement in the sale contract that a purchaser is being paid to clean up the land or that the purchase price is being reduced to reflect the contaminated state of the land.

Sellers/landlords can also avoid liability by selling with information – giving the purchaser (or tenant under a long lease) the necessary information to identify contamination before buying.

Where transactions have occurred since 1990 between large commercial organisations, the granting of permission by the seller for the buyer to carry out its own investigations as to the condition of the land is normally sufficient to indicate that the buyer had the necessary information.

Ground gas including radon
Trevor Rushton

Methane and carbon dioxide

Both methane and carbon dioxide gases occur naturally, but gas generation around areas of made ground, for example landfills, can give rise to dangerous concentrations. Radon is a naturally occurring radioactive gas that can be harmful if present in sufficient quantities.

Methane (chemical formula CH_4) is a basic hydrocarbon and is explosive in air at concentrations of between 5–15%. Methane is generated by the anaerobic (absence of oxygen) degradation of organic material.

Carbon dioxide (chemical formula CO_2) is a product of aerobic (presence of oxygen) organic degradation. Carbon dioxide is classed as highly toxic. Where 3%v/v carbon dioxide is present, this can result in headaches and shortness of breath, with increasing severity up to 5%v/v or 6%v/v. (Source: *Guidance on evaluation of Development proposals on site where Methane and Carbon Dioxide are present*, National House Building Council (NHBC), March 2007.)

Both methane and carbon dioxide are asphyxiants (methane by its capability to displace oxygen). Numerous studies have identified the risk presented by the entry of these gases into building voids and other confined spaces.

Perhaps the best known UK incident concerning landfill gas was at Loscoe, Derbyshire in March 1986, when a bungalow adjacent to a landfill site was destroyed by an explosion of accumulated methane. A subsequent investigation attributed a rapid drop in atmospheric pressure (27 millibars over 7 hours) resulting in gas being drawn in towards the property through the permeable underlying geology.

Another well-documented incident at Abbeystead in March 1984 resulted in 16 fatalities. Natural dissolved methane in water passing through a valve house at the Lune/Wyre water transfer scheme was released and allowed to accumulate during a period of inactivity. An unknown source (thought to be either an electrical fault or cigarette lighter) triggered the subsequent explosion, occurring at a time of an organised tour and demonstration of the valve house.

Sources

Natural sources of ground gas include:

- peat bogs and mosslands;
- limestone and chalk;
- coal measures;
- river and lake sediments;

- made ground;
- farmland; and
- sewers.

Detection and control

The assessment of risk from ground gas within new developments is a material consideration in local authority building control, planning, and environmental health legislation, including:

- The Town and Country Planning Act 1990 requires that the potential for contamination and risk from landfill and ground gases must be considered during development. If the development is within an area of potential risk, a planning condition will be likely to be attached to any permitted application, requiring satisfactory assessment and mitigation. The duty of the developer to maintain any gas protection systems may be a condition of the planning consent.

- Ground gas is also dealt with under Part IIA of the Environmental Protection Act 1990. Where a significant potential of significant harm exists to a development (either existing or being developed), the local authority can enforce appropriate remediation or mitigation.

- Approved guidance to the Building Regulations (Approved Document C, 2004 edition with 2013 amendments) states that where there is a potential risk, further investigation is required to determine whether gas measures are required and what level of protection is necessary.

A number of measures can be undertaken to reduce the risk of ground gas. A desk study may be undertaken to determine the probability of ground gas affecting the site. This includes a desk-based study of the history and geology of the area and any additional information such as mining or landfill activities. If a source of gas is identified, gas monitoring installations and monitoring may be recommended.

British Standard 10157:2001, *Code of Practice for investigation of potentially contaminated sites* and BS 5930:1999+A2:2010, *Code of Practice for site investigations* recommend a minimum of 12 months' monitoring.

The most common method of detection is a standard land-fill gas analyser and flow meter, which measure methane, carbon dioxide and oxygen concentrations. Photo ionisation detectors (PID) and flame ionisation detectors (FID) can be used where other gases, such as organic volatiles, are suspected.

Protection measures

The type of mitigation measures required depend on the type of development, gas type, volume, source, and emission rate.

In-ground barriers

These are physical barriers to block gas migration and usually comprise compacted clay, bentonite, or polyethylene. This method is often used for large areas, such as closed landfill sites. However, it is a relatively expensive protection measure.

Ventilation

Ventilation (either active or passive) is the most commonly used gas protection measure. A secondary level of protection is often required, as ventilation is influenced by factors such as wind speed and direction, and atmospheric pressure.

Ventilation can comprise underfloor venting or trench venting. Underfloor venting uses an open void space with connection to the atmosphere, such as air bricks, pipe riser, and gravel-filled trenches. Trench venting involves excavating a trench or series of columns and backfilling with a permeable infill such as gravel. Gas permeates freely to the open atmosphere rather than migrating laterally. This method is commonly used at the boundaries of landfill sites.

Gas resistant membranes

Membranes are often used in conjunction with other measures. A number of different types of membrane are available depending on the volatility and density of the gas. The effectiveness of a membrane is dependent on techniques of sealing the membrane to the building fabric, ensuring that gas cannot leak through to occupied areas or accumulate in voids.

Alarm systems

Alarm systems are designed to trigger when gas concentrations reach a certain level. Gas probes are situated around the building and feed data into a central system. These are mainly used as additional protection for high-risk sites or existing buildings where it is not possible to use other protection measures. They are, however, costly and require regular maintenance.

Radon

Radon is a colourless, odourless gas that is radioactive. It is formed where uranium and radium are present and can move through cracks and fissures in the subsoil, and so into the atmosphere or into spaces under and in dwellings. Where it occurs in high concentrations it can pose a risk to health (source: *Radon: Guidance on protective measures for new dwellings*, Building Research Establishment, 1999). The Health and Safety Executive (HSE) website notes that radon is now recognized as the second largest cause of lung cancer in the UK after smoking.

Radon is everywhere but usually in insignificant, variable concentrations. Problems occur when it enters enclosed spaces, such as basements, where concentrations can accumulate. Some areas in the UK are more exposed to radon due to localised geology.

The Health Protection Agency (HPA) publishes maps of radon-affected areas across England and Wales, Scotland, and areas of Northern Ireland. The HPA has set threshold levels for both commercial and residential properties. Where potential health risk may exist, detectors should be installed and remedial work undertaken to reduce exposure.

The UK reference site for Radon may be found at www.ukradon.org/

The British Geological Survey (www.bgs.ac.uk) also states whether radon is likely within a particular area as part of its Address-Linked Geological Inventory.

Legal requirements for workplaces – risk assessment

Under the Health and Safety at Work, etc. Act 1974, employers must, so far as is reasonably practicable, ensure the health and safety of employees and others who have access to their work environment. The Management of Health and Safety at Work Regulations 1999 require the assessment of health and safety risks.

Risk assessment for radon should be carried out in relation to:

- all below ground workplaces in the UK; and
- all workplaces located in radon affected areas.

Further guidance on risk assessments can be found at www.hse.gov.uk/radiation/ionising/radon.htm

Identification of radon

Radon levels can vary substantially with time, so prolonged measurements are required for reliable results. Short measurements can be misleading, low, or alarmingly high. The government recommends that people in affected areas test their property for a period of 3 months using passive monitors in order to provide a reliable estimate of the average radon level. Passive monitors are easy to use, inexpensive, and available from the Department for Environment, Food and Rural Affairs (Defra).

Remedial action for high radon levels can be quite straightforward. The best approach is to prevent radon entering the building from the ground by altering the balance of pressure between the inside and outside.

This can be achieved by carrying out the following:

- Install a small sump pump below the floor and connect to a low power fan in order to extract the air and reduce the pressure under the floor. Multiple sumps can be used in large buildings.

- Improved ventilation under suspended timber and concrete floors. New airbricks are installed, sometimes together with a fan.

- Increase the pressure in the building by blowing air (called 'positive pressurisation') from the roof space with a small fan. Best results are in buildings with low natural ventilation. Secondary benefits may include a reduction of other indoor pollutants such as carbon dioxide, reduced condensation, and a 'fresher' indoor environment.

- Alternatively, one may seal ducts, joints, and cracks in the floors, although this is rarely effective by itself and always laborious. It is helpful to close large openings when a sump is used.

- Install a membrane barrier. This is very difficult to successfully achieve in an existing building.

The BRE and the Health Protection Agency (HPA) have produced new guidance on reducing radon levels in the home, which focuses on improving underfloor ventilation to reduce domestic radon levels. The HPA recommends that householders with concentrations above the action level ($200\,Bq\,m^{-3}$) should take measures to reduce this, ideally to below the target level of $100\,Bq\,m^{-3}$.

GRG 37/1 *Radon solutions for homes – improving under floor ventilation* replaces BRE Report BR 270: *A guide to radon remedial measures in existing dwellings*. The guide is part one of a three-part set with parts two and three covering positive house ventilation and sump systems.

Radon and the Building Regulations

With the new understanding of radon risk, the government legislated that houses built since 1988 in parts of Devon and Cornwall and 1992 in parts of Somerset, Derbyshire, and Northamptonshire had to have radon protection measures built in. Additionally, the precise areas where radon protective measures should be taken are periodically reviewed by Defra as new data is provided by the Health Protection Agency.

Two zones of risk were allowed for. First the primary zone (the area with the highest risk). Requirement C2 of Schedule 1 of the Building Regulations requires that each house has a radon proof area, together with other precautionary measures that can be upgraded if a risk shows high radon levels. In the secondary zone (where the risk is lower) only precautionary measures must be built in. If a house has precautionary measures, upgrading them (e.g. adding a fan to a sump and pipe system) could solve the radon problem quickly and simply.

The Building Research Establishment (BRE) has published guidance on protective measures for new dwellings in support of the Building Regulations entitled BRE Report *Radon: Guidance on protective measures for new dwellings.* Equivalent advice has been prepared by the Health and Safety Executive (HSE) within a document entitled *Radon in the workplace.* The document recommends many similar measures are applicable for non-domestic buildings.

Electromagnetic fields

Trevor Rushton

'Electromagnetic fields' (EMFs) are a combination of electric fields and magnetic fields.

An electric field is created whenever a +VE or −VE electrical charge is present. A charged electrical wire will produce an electric field even if no current is flowing. The higher the potential difference (voltage), the stronger the field.

Magnetic fields are created when electric currents flow and therefore are not present unless an appliance is turned on. Magnetic fields are measured in nanotesla (nT), microtesla (mT), millitesla (mT) or tesla (T). Magnetic fields are not usually blocked by objects, although certain types of metal can shield against them.

Awareness of the health effects of electromagnetic fields (also known as electromagnetic radiation) has been steadily increasing since the 1970s as exposure to human-induced EMFs has increased both in the home and in the workplace, according to the World Health Organisation (WHO) 2007.

The Radiation Protection Division of the Health Protection Agency currently has the responsibility for providing advice on limiting exposure of the general public to EMFs.

What are the common sources of electromagnetic radiation?

EMFs are present everywhere in the environment, existing through the generate on and transmission of electricity, the use of electrical appliances and telecommunications (WHO, 2007). Of particular concern have been the effects of overhead and underground power lines, electricity substations, microwave ovens, computer and television screens, security devices, radars and the use of mobile phones.

Transmission power lines in the UK operate at 275 and 400 kilovolts (kV) and distribution lines at 440V, 11kV, 33kV, 66kV, and 132kV. The strength of EMFs decreases rapidly with increased distance from the source. Human epidemiology studies of magnetic fields have tended to use a field of 0.4 microtesla (μT) or above to identify potential risks. For high voltage power lines at 132kV and above, average field levels of 0.4μT or above may exist at ground level at distances greater than 100 metres. For lower voltage lines, a field of 0.4μT or above may occur up to a few tens of metres away.

EMFs can also occur naturally within the environment through the build-up of electrical charges associated with thunderstorms within the atmosphere.

What are the health effects associated with exposure to electromagnetic radiation?

A wide variety of symptoms and illnesses have been associated with exposure to EMFs, although these are not scientifically proven. These include:

- headaches;
- anxiety;
- suicide and depression;
- nausea;

- fatigue;
- premature pregnancies and low birth weight;
- cataracts;
- cardiovascular disorders;
- neurobehavioural effects and neurodegenerative disease;
- cancer, including childhood leukaemia;
- sleeping disorders; and
- convulsions and epilepsy.

Research and current knowledge

A large number of epidemiological studies have been undertaken across the world on the effects of exposure to EMF on human health. One of the most highly publicised issues is whether there is an increased risk of cancer within populations living near power lines or other sources of EMF. A great deal of research has been focused upon the risk of childhood cancer, in particular leukaemia.

Analysis of several studies has indicated that "the possibility exists of a doubling of the risk of leukaemia in children in homes at high levels of exposure to extremely low frequency (ELF) magnetic fields" (Health Protection Agency – Pooled Analysis, 2000). A task group set up by the WHO in October 2005 to assess risks to human health as a result of exposure to ELF EMFs concluded that additional studies since 2000 do not alter this view on the risk of childhood leukaemia.

The WHO task group in 2005 also assessed whether there was an association with ELF EMF exposure and the aforementioned associated health effects. It was concluded that 'scientific evidence supporting an association between ELF magnetic field exposure and all of these health effects is much weaker than for childhood leukaemia' (WHO, 2007).

The advisory group to the Health Protection Agency on non-ionising radiation has stated in relation to adults 'there is no reason to believe that residential exposure to EMFs is involved in the development of cancer'. However, it should be noted that there have been far fewer studies on the health effects on adults than on children.

What guidelines exist within the UK relating to EMFs?

A report published in 2004 by the former National Radiological Protection Board recommended the adoption in the UK of the guidelines produced by the International Commission on Non-Ionizing Radiation Protection (ICNIRP) for limiting exposures to EMFs having a frequency less than 300 GHz. The ICNIRP exposure guidelines relate to occupational and public exposure to EMFs.

Research into the effects of EMFs is under constant review by the Health Protection Agency to ensure guidelines are in line with the most up to date information. Guidelines are therefore susceptible to change. For the most up to date guidelines and information, the Health Protection Agency (while still in existence) and ICNIRP websites should be consulted.

Further information

World Health Organisation, June 2007, *Electromagnetic fields and public health – Exposure to extremely low frequency fields*, Fact Sheet No. 322: www.who.int/media-centre/factsheets/fs322/en/print.html.

National Radiological Protection Board, 2004, *Advice on limiting exposure to electromagnetic fields (0–300 GHz)*, Documents of the NRPB, Volume 15 No. 2.

Health Protection Agency website: www.hpa.org.uk/.

International Commission on Non-Ionizing Radiation Protection website: www.icnirp.de/.

Ecology and bat surveys
Michael Wright

Consideration of ecology and biodiversity in development or demolition

Some animal species are protected from disturbance by virtue of their vulnerability such as bats. However, legislation also exists to protect, maintain, and encourage biodiversity, and this will inevitably encompass many different species of plants and/or animals.

Proposed developments that require planning permission, particularly larger schemes, are more likely to require consideration for the protection of any existing animal inhabitants in and around a building. Protected species include bats and other species, such as protected birds, including peregrine falcons, owls, and the black redstart. Bats are the most common inhabitants of our buildings and as such are covered in more detail within the next section.

Bats

Bats account for nearly a quarter of all mammal population in the UK, and their existence is protected by law. The reduction in the number of natural roosting sites across the country has led to more and more bats forming roosts in the man-made structures which, when maintained or refurbished, can threaten the security of roosts. There are 17 species of bat in Britain and Ireland. All are found in the south of England, but the number of species declines further north, with only five in northern Scotland. All are protected by law, and it is a criminal offence to deliberately kill, injure, disturb, or capture a bat, or to damage or destroy their roosts. It is important to note that roosts are still protected even when the bats are not physically present. Offences attract fines of up to

£5,000 or 6 months in jail per bat. The overriding legislation protecting bats is the Wildlife and Countryside Act 1981 and The Conservation (Natural Habitats &c.) Regulations 1994.

Prior to any major refurbishment or demolition surveyors need to be aware of the implications of finding bats or other species within these structures. Changes such as upgrading or providing artificial lighting can affect bats' navigation senses and could cause bats to avoid or desert a roost and/or affect their emergence times.

Bat surveys

Surveyors involved with the management, maintenance, and refurbishment of properties, particularly traditional properties, should be aware of the obligations in relation to bats or other species that may be present. If you do not carry out a survey and bats are discovered in the course of the works, the works are likely to be halted until a suitable scheme of work has been agreed.

The following five-step process as advocated by English Heritage, The National Trust, and Natural England may help the surveyor to comply with legislation where works are proposed to a building where bats are roosting:

1 Contact Natural England, who may send a licensed volunteer bat worker to your building to assess the situation. Some works may be permissible without a licence.
2 Appoint a bat consultant to help ensure legislation is being complied with.
3 Commission a survey to establish what species of bats are present and their numbers. Any licence application will require this information.
4 Obtain a report to illustrate where the bats are, and whether they will be disturbed during the works.

5 Apply for a licence with the help of your bat consultant in support of the proposed works, and if successful, you will be granted a licence.

Further information
The Bat Conservation Trust website contains useful information on bats and free pdf publications on *Bat surveys – Good practice guidelines, bats and lighting in the UK: www.bats.org.uk/*.

BRE Good Repair Guide GRG 36: *Bats and refurbishment*, BRE, May 2009.

National Building Specification Shortcuts: NBS Shortcut 76, *Do not Disturb: Bats in buildings*, June 2009.

Planning Policy Statement 9: Biodiversity and Geological Conservation (PPS9), 2005. Bats and other species are given material consideration during the planning process. Measures to promote or protect species (such as bat or bird boxes) may be specified by authorities.

Bats in Traditional Buildings 2009. English Heritage, National Trust and Natural England.

5.2 Sustainable development

Code for sustainable homes
Philip Chorley

Driven by the wider requirements of the EU Energy Performance of Buildings Directive, the Code for Sustainable Homes (CSH) has replaced previous EcoHomes requirements to become the single national standard to measure sustainability of **new** homes. On 30 May 2014, the government published an addendum to bring the Code for Sustainable

Homes (CSH) into line with recent regulatory and national guidance. If regulatory compliance is to SAP 2012, then the new 2014 CSH addendum should be followed with evidence based on the SAP 2012 calculations. If the site was registered under part L 2010, then the code assessment should be registered to the November 2010 guidance with evidence to this effect. Any residential refurbishments can be certified under BREEAM Domestic Refurbishment Standard (a voluntary scheme).

The Code for Sustainable Homes was launched on 13 December 2006 and since April 2012 it has completely replaced EcoHomes (except in Scotland). EcoHomes has also been replaced for residential refurbishments with the BREEAM Domestic Refurbishment Standard which was issued in July 2012, the most recent edition of the code being released in October 2014.

CSH has been used by the Homes and Communities Agency (HCA) in England, Wales, and Northern Ireland, where it is mandatory to build affordable homes to at least Code Level 3.

Although this Code is produced by BRE, it is 'owned' by the Department for Communities and Local Government (DCLG) (in contrast to the other BREEAM systems), so the designation of how the credits are applied is dictated by DCLG.

Assessments must be undertaken by a Licensed Code Assessor.

Assessment involves a pre-assessment at the planning stage; and then formal assessments are made at the design and post-construction stages (which include a BRE audit), after which a certificate is issued.

A list of qualified assessors can be obtained from BRE (see www.breeam.org).

How to score credits

Under the Code, each area or category is broken down into a number of 'issues'. Credits are awarded against these issues, and then a weighting is applied to those credits scored in each category (see the Code Categories table later) to get the final code score (total number of points). The Technical Guide should be referred to for the detail (and the most up to date description), as new editions are constantly being issued. The latest at going to print is Nov 2010, which has tried to align the Code with Part L (2010) (with 2011 amendments; to view the document, see www.planningportal.gov.uk/uploads/code_for_sustainable_homes_techguide.pdf).

Code levels

The Code has six levels (see the following table):

Table 5.1

Code levels	Description	Total score
Level 1 (★)	above regulatory standards and a similar standard to BRE's EcoHomes PASS level and the Energy Saving Trust's (EST) Good Practice Standard for energy efficiency	36 points
Level 2 (★★)	a similar standard to BRE's EcoHomes GOOD level	48 points
Level 3 (★★★)	a *broadly* similar standard to BRE's EcoHomes VERY GOOD level and the EST's Best Practice Standard for energy efficiency	57 points
Level 4 (★★★★)	broadly set at current exemplary performance Code level 4 is a 44% improvement above Part L 2006 and 25% above Part L 2010.	68 points

Code levels	Description	Total score
Level 5 (★★★★★)	based on exemplary performance with high standards of energy and water efficiency	84 points
Level 6 (★★★★★★)	aspirational standard based on zero carbon emissions for the dwelling and high performance across all environmental categories	90 points

Conclusion

One hopes that the Code will offer a range of benefits, in terms of:

- the reduced environmental impact of new homes provided;
- to consumers, through lower fuel costs; and
- to house builders, as a mark of quality assurance.

Further information
For access to the Technical Guidance and other useful information go to:

www.breeam.org/page.jsp?id=86
www.planningportal.gov.uk/buildingregulations/greener
buildings/sustainablehomes
http://www.breeam.com/page.jsp?id=228
www.planningportal.gov.uk/buildingregulations/greener
buildings/sustainablehomes

Building information modelling

Tom Dewhurst

What is building information modelling (BIM)?

> BIM is described as being 'a digital representation' of physical and functional characteristics of a facility creating a shared knowledge resource for information about it forming a reliable basis for decisions during its lifecycle, from earliest conception to demolition.
>
> *(Source: BIM in the UK, NBS briefing January 2011, RIBA et al.)*

BIM finds its roots throughout the construction industry, including the principle of Partnering which was introduced in the 1990s following Latham's 'Constructing the team' report in 1994. Through evolving government targets, improvements in technology and sustainable agendas, the use of BIM has become a focus within the construction industry in the past decade.

BIM allows the collaboration and sharing of asset information between those in the construction industry, in order to provide significant improvements in cost, value and carbon performance. BIM goes far beyond the use of technology. It allows 3D, 4D (time), and 5D (cost) modelling to be incorporated into one virtual building model, facilitating design, feasibility testing, and analysis by different parties on the same model at the same time, thus reducing the need for reproduction of design information.

The model also contains information about the construction and finishes of each element, and is able to store information on the resources needed for construction, enabling accurate specification and programming to be produced. The database of information created can be used throughout the construction process from inception, with the use of clash detection, right through to post occupancy, where facilities managers are able to use the model to better manage the building. It is therefore more about the database of information than it is about the visual model itself.

BIM and its benefits

Because all parties collaborate at an early stage in the design, on-site alterations are reduced dramatically. This can lead to a reduction in waste, estimated at between 20–30% of a project. This is possible as the project can be fully constructed virtually, with any changes in design implemented at each stage, allowing any snags or design faults to be eliminated with input from members from all disciplines.

BIM can also allow greater cost certainty for the client. As the model is fully constructed virtually, there are less costly onsite adjustments and therefore the project budget is less likely to be exceeded. This means less risk for the client. The software allows input from manufacturers, suppliers, and cost consultants allowing, subconsultants to more accurately cost their works using the information within the model.

BIM and its problems

The real problem with BIM revolves around the general inertia within the industry to adopt this new way of working. BIM, by its very nature, relies on the full collaboration of all parties involved, without this the project falls down.

Despite recent advancements in technology there are still problems with the transfer of data, essential to the success of BIM, between the various types of software. A good example of this is the use of COBIE data (Construction Operations and Building Information Exchange). COBIE data is essentially a large database of all the information which is created within the model. It was created in the USA and as such is not compliant with most software in use in the UK today requiring extensive code to be written in order to create a link between the model and the spreadsheet.

Another drawback of using BIM revolves around the perception of liability between the different parties involved as everyone is working on the same central model. This is, in part, overcome by the use of software originator, read and write permissions alongside shared risk procurement routes such as partnering.

Despite a slow uptake there have now been several new contract types developed to support the challenges that this type of collaborative working pose. Consensus DOC 300 and the Integrated Form of Agreement (IFOA) both create a contractual and financial framework to facilitate the collaboration between construction project participants. There are also changes within the NEC 3 suite of contracts released in April 2013. These changes allow provisions for additional clauses to be added if required to facilitate the implementation of BIM. As part of the NEC3 documents there has also been a how-to guide created for use on BIM projects.

BIM software

Currently there is no one piece of software available to encompass all of the building functions, therefore various programmes will contribute to BIM. It is essential that there is a common language between differing software packages. Developed by the Building Smart Organisation, a standard language known as IFC (Industry Foundation Classes) has been created to facilitate this. International standard (ISO 16739) is the industry standard for IFC.

BSI has produced a standard (ISO 29481-1) for the production of an Information Delivery Manual (IDM). The standard gives guidance for production and issuing of documents (see www.bsigroup.com/standards).

Autodesk has a number of software packages available for BIM, the most common of which is Revit (see http://usa.autodesk.com).

Implementation of BIM

There are currently five recognised levels to BIM:

- Level 0 is no adoption (simple CAD drawings).
- Level 1 is a mix of 3D CAD for design and 2D for sharing documentation.
- Level 2 is the target set by the Government for 2016 and involves collaborative working where all parties are using their own 3D CAD models with information being exchanged between the parties.
- Level 3, referred to as the 'holy grail' by the NBS, represents full collaboration between all parties who are all using one shared model held in a central repository. This is referred to as 'Open BIM' and is the government target for all public sector work by 2019.
- Looking into the future there may be a further level added which involves the use of 4D BIM involving BIM data to analyse time, cost, and facilities management.

Careful consideration will need to be given when devising a strategy for adoption of BIM. Using a pilot project, which simulates real development, can mitigate expensive mistakes and allows the full use of support and training offered by software vendors. The costs of implementing BIM have widely been estimated at £25,000 per operator for the software and all training required with a 6-year licence. The capital expenditure for implementation is not the major investment. The time required for training staff and changing their way of working is what creates issues when implementing BIM.

Despite the initial costs involved with implementing BIM, a study produced for the BSI investors report in 2010 studied five key projects utilising BIM and highlighted between an 8–18% saving in design fees, 8–10% saving on pre-construction phase costs, and a further 8–10% saving during the construction stage.

A study by the RICS in 2013 stated that the lack of client demand was the largest barrier to the uptake of BIM. As the understanding of the benefits become more widely appreciated within the industry the requirement for BIM will undoubtedly increase. Alan Muse, director of the built environment professional group at the RICS said:

> Education will be critical to initiating the cyclical change needed here – leading to increased practical implementation of BIM, greater recognition of the benefits it can bring, and ultimately heightened demand for its usage.

Formulation of a strategy for implementation of BIM should consider the following:

- nomination of a senior person to champion BIM;
- selection of team members who are enthusiastic about BIM and are up for the challenge;
- identification of a training programme which complements organisational objectives;
- the level of support services offered by the software vendor;
- required changes to existing processes; and
- direct and indirect costs associated with implementation of BIM.

The future for BIM

BIM is not a new concept and it is already being effectively used on new build and refurbishment schemes with equal success. The government targets for 2016 and 2019 will inevitably drive its implementation in the public sector eventually having a knock-on effect for those working in the private sector too.

Further information

www.constructingexcellence.org.uk
www.bimtaskgroup.org/bim-regional-hubs
www.bre.co.uk/bim

6
Useful resources

The RICS Assessment of Professional Competence
Alistair Cooper

The Assessment of Professional Competence (APC) is practical training and experience which, combined with academic qualifications, leads to professional membership of the Royal Institution of Chartered Surveyors (RICS). There are 20 APC pathways, which relate to land, property, construction, and the environment.

Successfully completing the APC entitles RICS members to practise as qualified chartered surveyors. There are a number of routes to membership and details of these, the APC pathways and specific requirements are set out on the RICS website at www.rics.org/uk/apc/.

What does it involve?

Candidates are required to complete a period of structured training. Depending upon their entry route, this will normally be for a minimum of 23 months, with a minimum of 400 days' relevant practical experience. During this period they are required to appoint an APC counsellor and supervisor, who will provide guidance and support on their training and day-to-day work. It is recommended, although not mandatory, that a candidate's counsellor and supervisor work at the same company, enabling day-to-day guidance. The counsellor must be a chartered surveyor.

The APC training is competency-based and candidates are required to demonstrate that they have the skills and abilities to satisfy specific competencies.

Assessment takes the form of written submissions; these are composed of an appraisal of a suitable project which the candidate has been involved with during their training and detailed accounts of their knowledge and experience under each competency. Candidates are required to attend a professional interview, known as the 'final assessment', involving a presentation to a panel of RICS assessors and answering detailed questions relating to their submission documents and broader aspects of their experience and knowledge.

RICS has produced a series of guides and templates to assist candidates, supervisors, counsellors, and employers, all of which can be downloaded from the RICS website and in particular, the APC guides provide essential advice on the competency requirements for relevant pathways.

Ten APC tips

1 **Choose your employer wisely**: Attempt to gauge a firm's commitment to APC training by enquiring about the level of support and training offered, along with reviewing their structured training agreement. It is also essential to gain the correct spread of work experience during your training period to satisfy relevant competency requirements.

2 **Get organised**: The timing of your APC application and other submissions can be critical. For example, a delay in your enrolment could hold up your final assessment by 6 months. Clarify requirements and diarise key dates and milestones. Final assessment interviews for building surveyors are held twice a year, usually during May and November.

3 **Carefully consider competency choice**: You should do some research and set about gaining good advice with

regard to competency choices from your employer and colleagues who have recently undertaken their APC. It is essential to ensure that there is a 'fit' with the range and nature of work you will be undertaking.

4 **Get out and about**: There is no substitute for quality one-to-one on-the-job training; so try to ensure that you make the most of shadowing opportunities from the outset of your training period. If your employer is multi-disciplinary try to gain experience with other disciplines, such as project managers and quantity surveyors. This experience will broaden your knowledge, thus aiding you in the final assessment.

5 **Take ownership of your training**: It is your APC. Make time to administer your APC paperwork and proactively set about identifying and planning your own training needs and work experience requirements. Your supervisor and counsellor should support and guide you; however, it is ultimately down to you to ensure you gain the correct experience.

6 **Network and get advice**: Join in with company activities, speak to colleagues about their experience of the APC, consult with your local RICS APC Mentor, and get involved with RICS MATRICS, the organisation for young chartered surveyors, trainees, graduates, and friends you graduated with now working in the profession. You will benefit from advice, broaden your perspective of the profession, and, as importantly, have fun and make lifelong friendships.

7 **Professional approach**: Keeping abreast of current trends and hot topics is essential. Reading relevant property and industry magazines and keeping up to date with changes in legislation, best practices, and the like will help build your confidence as a rounded professional, along with providing you hours to log for Continuous Professional Development (CPD).

8 **Personal development**: You will need to develop personal skills during the course of your training period in preparation for the final assessment. Good time manage-

ment, organisation, and presentation skills are essential. These cannot be attained at the last minute.

9 **Technical competence**: Continually develop your technical ability. Be resourceful, grow interest in your subject area, attend seminars, present to your colleagues and undertake structured reading relevant to your competencies.

10 **Prepare for the final assessment**: Allow plenty of time to prepare for the final assessment interview. Ask experienced colleagues to conduct mock final assessment interviews, ideally prior to submitting. Receive feedback, refine your presentation and repeat as necessary. Attempt to gain interview experience with other companies.

Further information

Royal Institution of Chartered Surveyors (RICS)
Website: www.rics.org/uk/apc and www.rics.org/uk/join
Email: contactrics@rics.org
Telephone: +44 (0)24 7686 8433

Watts Group Limited
Website: www.watts.co.uk/working-for-us
Email human resources: human.resources@watts.co.uk
Telephone: +44 (0)20 7280 8000

Useful references
These guides are all available to download free on the RICS website:

APC candidates guide;
APC requirements and competencies guide;
APC guide for supervisors, counsellors, and employers;
APC pathway guides; and
The RICS global professional and ethical standards

Conversion factors

Trevor Rushton

Within the following table, to convert the units in Column A into the equivalent in Column B, multiply by the factor shown. To convert the units in Column B into the equivalent in Column A, divide by the factor shown.

Table 6.1

Units in Column A	Factor	To give units in Column B
acres	0.404685642	hectares
acres	4046.8627	m^2
atmospheres (atm)	0.007348	ton/sq inch
atm	76	cm of mercury
atm	14.7	pounds/sq in.
atm	1,058	tons/sq ft
bars	10,200	kgm^{-2}
bars	2,089	pounds/sq ft
bars	14.5	pounds/sq in.
Btu	10.409	litre-atmosphere
Btu	1,054.8	joules
Btu	0.0002928	kWh
Btu/hr	0.2931	Watts (W)
cm	0.03281	feet
cm	0.3937	inches
cm	0.00001	km
cm	0.01	m
cm	10	mm
cm^2	0.001076	sq feet
cm^2	0.1550	sq inches
cm^2	0.0001	m^2
cm^2	100	mm^2
cm^2	0.0001196	sq yards
cm^3	0.0003531	cu feet
cm^3	0.06102	cu inches

Units in Column A	Factor	To give units in Column B
cm³	0.000001	m³
cm³	0.000001308	cu yards
cm³	0.0002642	gallons (US liq.)
cm³	0.001	litres
cm³	0.002113	pints (US liq.)
cm³	0.001057	quarts (US liq.)
cubic feet	28,320	cm³
cubic feet	1,728	cu inches
cubic feet	0.02832	m³
cubic feet	0.03704	cu yards
cubic feet	7.48052	gallons (US liq.)
cubic feet	28.32	litres
cubic feet	59.84	pints (US liq.)
cubic feet	29.92	quarts (US liq.)
cubic inches	16.39	cm³
cubic inches	0.0005787	cu feet
cubic inches	0.00001639	m³
cubic inches	0.00002143	cu yards
cubic inches	0.004329	gallons (US liq.)
cubic inches	0.01639	litres
cubic inches	0.03463	pints (US liq.)
cubic inches	0.01732	quarts (US liq.)
cubic inches	28.38	bushels (dry)
cubic yards	764,600	cm³
cubic yards	27	cu feet
cubic yards	46,656	cu inches
cubic yards	0.7646	m³
cubic yards	202	gallons (US liq.)
cubic yards	764.6	litres
cubic yards	1,615.9	pints (US liq.)
cubic yards	807.9	quarts (US liq.)
cubic yards/min	0.45	cubic ft/sec
cubic yards/min	3.367	gallons/sec
cubic yards/min	12.74	litres/sec
days	86,400	seconds

Units in Column A	Factor	To give units in Column B
drams	1.7718	grammes
drams	0.0625	ounces
feet	30.48	cm
feet	0.0003048	km
feet	0.3048	m
feet	0.0001645	miles (naut.)
feet	0.0001984	miles (stat.)
feet	304.8	mm
feet/sec	30.48	cm/sec
feet/sec	1.097	km/hr
feet/sec	0.5921	knots
feet/sec	18.29	m/min
feet/sec	0.6818	miles/hr
feet/sec	0.01136	miles/min
gallons	0.004951	cu yards
gallons	3.785	litres
gallons (liq. Br. Imp.)	1.20095	gallons (US liq.)
gallons (US liq.)	0.83267	gallons (Br. Imp.)
gallons of water (US liq.) (US)	8.3453	pounds of water
gallons of water (liq. Br. Imp.)	10.022	pounds of water
gallons/min (US liq.)	0.0002228	cu ft/sec
gallons/min (US liq.)	0.06308	litres/sec
gallons/min (US liq.)	8.0208	cu ft/hr
gallons/min (liq. Br. Imp.)	4.54609	litres/min
Gallo(US liq.)ns/sec	0.8326725	Gallons (Br. Imp.)/sec
grammes	0.001	kg
grammes	1,000	mg
grammes	0.03527	ounces (avdp)
grammes	0.03215	ounces (troy)
grammes	0.002205	pounds
hectares	2.471053816	acres
hectares	10,000	m^2
horsepower	33,000	foot-lbs/min

Units in Column A	Factor	To give units in Column B
horsepower	550	foot-lbs/sec
horsepower	10/68	kg-calories/min
horsepower	0.7457	Kilowatts (kW)
horsepower	745.7	watts
inches	2.540	cm
inches	0.002540	m
inches	25.40	mm
joules	0.0009480	Btu
kg	1,000	grammes
kg	9.807	joules/metre (newtons)
kg	2.205	pounds
kgm^{-2}	0.00009678	atmospheres
kgm^{-2}	0.00009807	bars
kgm^{-2}	0.003281	feet of water
kgm^{-2}	0.002896	inches of mercury
kgm^{-2}	0.2048	pounds/sq ft
km	3,281	feet
km	39,370	inches
km	0.6214	miles
km/hr	27.78	cm/sec
km/hr	54.68	feet/min
km/hr	0.9113	feet/sec
km/hr	0.5396	knots
km/hr	16.67	m/min
km/hr	0.6214	miles/hr
kWh	3,413	Btu
kw	56.92	Btu/min
kw	1,000	watts
litres	1,000	cm^3
litres	0.03531	cu feet
litres	61.02	cu inches
litres	0.001	m^3
litres	0.001308	cu yards
litres	0.2642	gallons (US liq.)
litres	2.113	pints (US liq.)

Units in Column A	Factor	To give units in Column B
litres	1.057	quarts (US liq.)
m	0.0005396	miles (naut.)
m	0.0006214	miles (stat.)
m	1,000	mm
m	1.094	yards
m	3.2808	feet
m^2	0.000247105	acres
m^2	0.0001	hectares
m^2	10.7639104	square feet
m^3	1,000,000	cm^3
m^3	35.31	cu feet
m^3	61,023	cu inches
m^3	1.308	cu yards
m^3	264.2	gallons (US liq.)
m^3	1,000	litres
m^3	2,113	pints (US liq.)
m^3	1,057	quarts (US liq.)
metres per min (m/min)	1.667	cm/sec
m/min	3.281	feet/min
m/min	0.05468	feet/sec
m/min	0.06	km/hr
m/min	0.03238	knots
m/min	0.03728	miles/hr
m/sec	196.8	feet/min
m/sec	3.281	feet/sec
m/sec	3.6	km/hr
m/sec	0.06	km/min
m/sec	2.237	miles/hr
m/sec	0.03728	miles/min
miles (statute)	5,280	feet
miles (statute)	63,360	inches
miles (statute)	1.609	km
miles (statute)	1,609	metres (m)
miles (statute)	0.868357	miles (nautical)
miles (statute)	1,760	yards

Units in Column A	Factor	To give units in Column B
miles/hr	44.70	cm/sec
miles/hr	88	feet/min
miles/hr	1.467	feet/sec
miles/hr	1.609	km/hr
miles/hr	0.02682	km/min
miles/hr	0.8684	knots
miles/hr	26.82	m/min
miles/hr	0.1667	miles/min
ml	0.001	litres
mm	0.1	cm
mm	0.003281	feet
mm	0.03937	inches
mm	0.000001	km
mm	0.001	m
ounces	0.0625	pounds
ounces	0.9115	ounces (troy)
pounds/cu ft	0.01602	grammes/cm^3
pounds/cu ft	16.02	kg/m^{-3}
pounds/sq in.	703.1	kgm^{-2}
pounds/sq in.	144	pounds/sq ft
quarts	0.9463	litres
square feet	0.00002296	acres
square feet	929	cm^2
square feet	144	sq inches
square feet	0.0929	m^2
square feet	0.1111	sq yards
watts	3.413	Btu/hr
watts	0.05688	Btu/min
watts	0.001341	horsepower
watts	0.00136	horsepower (metric)
watts	0.01433	kg-calories/min
watts	0.001	kilowatts

To convert temperature

$°C = 5/9 \ (°F) - 32$

$°F = 9/5 \ (°C) + 32$

Watts.

BUILDING
RELATIONSHIPS.

Watts Group Directory

Property and Construction Consultants
watts.co.uk

Belfast
Wellington Buildings
2–4 Wellington Street
Belfast
BT1 6HT
T: +44 (0)28 9024 8222

Birmingham
Albert Wing, The Argent Centre
60 Frederick Street
Birmingham
B1 3HS
T: +44 (0)121 265 2310

Bristol
25 Marsh Street
Bristol
BS1 4AQ
T: +44 (0)117 927 5800

Edinburgh
10 Castle Street
Edinburgh
EH2 3AT
T: +44 (0)131 226 9250

Glasgow
Centrum Building
38 Queen Street
Glasgow
G1 3DX
T: +44 (0)141 353 2211

Leeds
49a St Paul's Street
Leeds
LS1 2TE
T: +44 (0)113 245 3555

London
1 Great Tower Street
London
EC3R 5AA
T: +44 (0)20 7280 8000

Manchester
St James's Tower
7 Charlotte Street
Manchester
M1 4DZ
T: +44 (0)161 831 6180

 @Watts_Group

 Watts Group Limited

Index